BOOK · 6

THE
STABLE
YARD

THE BRITISH HORSE SOCIETY

THE MANUAL OF
STABLE MANAGEMENT

COMPILED BY
Pat Smallwood FBHS

THE ADVISORY PANEL INCLUDED
Barbara Slane Fleming FBHS
Tessa Martin-Bird FBHS
Stewart Hastie MRCVS
Jeremy Houghton-Brown
E.E. Ellis
Helen Webber FBHS

SERIES EDITOR
Jane Kidd

Reprinted in enlarged format 1992

© British Horse Society 1989

British Horse Society manual of stable
    management
    Bk.6: The Stable Yard
    1. Livestock: Horses. Management
    I. British Horse Society, *Advisory Panel*
    636.1'083

    ISBN 1-872082-28-9

Produced for The British Horse Society by
The Kenilworth Press Limited,
Addington, Buckingham, MK18 2JR

Typeset by Rapid Communications Ltd
Printed and bound in Great Britain by
Hollen Street Press Ltd, Slough, Berks.

CONTENTS

Introduction

The aim of this series is to provide a reliable source of information and advice on all practical aspects of horse and stable management. Throughout the series emphasis is placed on the adoption of correct and safe procedures for the welfare of all who come into contact with horses, as well for the animals themselves.

The books have been compiled by a panel of experts, each drawing on considerable experience and contributing specialised knowledge on his or her chosen subject.

The other titles in the series are:

Book 1: The Horse – Conformation; Action; Psychology of the Horse; Teeth and Ageing; Breeds; Breeding; Identification; Buying and Selling; Glossary of Terms.

Book 2: Care of the Horse – Handling the Horse; Stable Vices and Problem Behaviour; Grooming; Bedding; Clipping, Trimming, Pulling and Plaiting; Recognising Good Health and Caring for the Sick Horse; Internal Parasites; Shoeing.

Book 3: The Horse at Grass – Grassland Management; Management of Horses and Ponies at Grass; Working the Grass-kept Pony or Horse; Bringing a Horse up from Grass.

Book 4: Saddlery – Saddles; Bridles; Other Saddlery; Bits; Boots and Bandages; Clothing; Care and Cleaning of Leather; Saddling and Unsaddling.

Book 5: Specialist Care of the Competition Horse – Dressage Horse; Driving Horse; Show Jumper; Event Horse; Long-Distance Horse; Hunter; Show Horse or Pony; Point-to-Pointer; Polo Pony; Types of Transportation; Travelling.

Book 7: Watering and Feeding – Watering; Natural Feeding; The Digestive System; Principles of Feeding; Foodstuffs; Rations; Problem Eaters; The Feedshed, Storage and Bulk Purchasing.

NOTE: *Throughout the book the term 'horses' is used and it will often include ponies.*

CHAPTER 1
Construction of Stables

Before constructing a new stable yard:

1 Consider your financial situation.
2 Obtain outline planning permission from the local authorities.

SITE

Any local byelaws relating to drainage and construction should be checked. If there is a choice of site, consideration should be given to the following:

☐ Accessibility. If an access road has to be built, this will increase the overall cost of the project. Installing services such as water, electricity and telephone, even over short distances, can be very expensive.

☐ Drainage. Stabling requires a solid, dry, well-drained base. If the soil is sand, gravel or chalk, this should present no problem. With clay or a high water level, the sub-soil has to be removed, replaced, and built up with hard core and gravel.

☐ Convenience. Ideally the stables should be sited so that they are visible from the house or staff living quarters. Regular checks can then easily be made.

YARD AND ACCESS ROAD

Foundation

A good foundation is essential. If building a new stable yard, the access road may be put down first to give a firm base for lorries. It is usual to excavate down to the sub-soil, but should it be an area of poor drainage and clay, it may be necessary to go deeper and remove the underneath soil. The whole area is back-filled with hard core or similar material according to availability locally. It is compacted with a vibratory roller.

These foundations should extend at least about 1m (3ft) beyond the area of the yard and 60cms (2ft) on either side of the access road.

The water main and electric cable which come underground from the public road or neighbouring house and buildings should be put in before the top surface is put on. They should emerge in the tack room or office area. If the water main is to cross the yard where heavy lorries will manoeuvre, it is likely to be fractured by pressure. To avoid this and to prevent the water freezing in cold weather, the main should be laid in a 60cm (2ft) deep trench to the side of the roadway.

The roadway should be edged with a concrete kerb, to prevent the sides from being undermined by surface water and giving way under pressure. The access road should have a width of 3.6m (12ft), and preferably should open out to a distance of 12m (40ft) as it reaches the road. This entrance should be surfaced to a width of 6m (20ft); to save expense the remainder may be left. The 12m (40ft) clear expanse should give good visibility for both riders and traffic.

Drainage

Surface water off the yard will collect at the lowest point, and should be directed off and away from the buildings. If the original area is flat, it must be arranged that the top surface should have a slight slope approximately one in 60 towards an open shallow drain. This surface water can then be directed:

(a) Away across the field by means of a 'French drain';
(b) To a soak-away; or
(c) If (a) and (b) are not possible, then to a septic-tank system installed to take the stable drains.

A 'French drain' is a ditch of varying lengths dug to a suitable depth according to requirements, shallower at the yard and deeper as it gets to the field. It is filled with reject stone.

A soak-away is a large circular hole 3 to 6m (10 to 20ft) deep, filled with reject stone. This works successfully in a sand, gravel or chalk area, but should not be considered in a clay district.

If surface water is to join a drainage system, then one or more gullies will have to be constructed to collect surface mud and straw. They must have strong, removable steel covers, and be large enough to cope with heavy rainwater.

YARD AND ROAD SURFACES

Types
(a) Gravel, stones or chippings are attractive to look at and non-slippery, but are difficult to keep clean and tidy, and are not suitable for a yard with large numbers of horses.

(b) Concrete and tarmac are practical, although tarmac is inclined to be slippery. They are relatively easy to keep clean and tidy, and are of good appearance. If put down of a suitable thickness they will last for many years, providing they have good drainage. Frost can be damaging to concrete and tarmac. Also, they are slippery in icy weather, and suitable precautions have to be taken, i.e. salt or sand put down, or straw tracks made.

The access road can be left untopped, but will then eventually break up. A top surface of concrete or tarmac makes a sound, long-lasting surface, and enhances the approach to the yard.

All yard and road surfaces need to be well maintained, and repaired as necessary. If cracked or broken surfaces are left, the deterioration becomes more rapid and more expensive to put right.

All yards must have a good gate and fences to prevent livestock getting on to the public highway. The gate must be wide enough for easy access, well hinged and have a sound fastening device. The yard gate should be kept shut at all times.

SHELTER

If a barn-type stable is selected (see *Types of Stabling*, page 12), shelter from cold winds need not be considered. If conventional lines of loose boxes are planned, try to select a sheltered area protected by other buildings, or by hills or trees. If building in a high, exposed position: an 'L' shaped design, three sides of a square, or a rectangle, will give more protection.

New Ideas

Research is being carried out with regard to flooring, ventilation, and labour saving. Anyone considering the building of new stables should study carefully all the various possibilities before going ahead with any plans.

Design Considerations

TACK ROOM AND FEED SHED. These should be near the stable-block, and preferably should be connected by an overhang. As the tack room is a security risk, it should be in view of the staff living quarters.

OFFICE, LAVATORIES, CHANGING ROOMS, STAFF AND LECTURE ROOMS should be adjacent to each other and with access to the car park.

MUCK HEAP. This should be sited away from the car park

and the yard entrance, but should be within easy reach of the stables. The road leading to it should have a surface firm enough to bear heavy lorries. The muck heap should cover an area which is practical for the size of the particular yard. It should have a concrete base and there should be a concrete path leading to the stables. The surrounds should slope so that the surface water will flow to an open drain and will be channelled away – but it should not be directed across fields because of the worm load · One side should be open and the other three sides built up to a height of 1.8m (6ft). Railway sleepers or concrete blocks can be used for this purpose · The path can be raised or the muck heap dug out so that manure can be thrown down rather than up.

HAY BARN. This should not be adjacent to the stabling because of the risk of fire. Preferably it should be connected to the yard by a hard surface wide enough to accommodate a large hay lorry.

POWER LINES. If these are above ground, they must be high enough to give clearance for muck and hay lorries.

YARD SURFACE. This can be of tarmac or concrete – but remember that all surfaces other than gravel are slippery in icy weather. Gravel is not practical if there are large numbers of horses in the stables.

CHAPTER 2

Design of Stables

The design of stabling will be determined by:

☐ Planning authorities.
☐ Local conditions: e.g. lie of the land, shelter, drainage and other buildings.
☐ Personal preference.
☐ Cost.

TYPES OF STABLING

Individual Loose Boxes

These are the most popular choice. Each stable is complete in itself and can stand alone or be built in (a) parallel lines, (b) an 'L' shape, (c) a square or (d) an open-ended square. Types (c) and (d) are considered to be more liable to facilitate the spread of disease, as there is less circulation of fresh air. On the other hand, on exposed sites they give more protection. An overhang should be incorporated along the front of the boxes to give shelter to horses and grooms.

American Barn

Barn stabling consists of a large covered area containing two rows of loose boxes facing each other across a 2.5 to 3.7m (8 to 12ft) passageway.

Advantages
☐ Pleasant working conditions for the staff.

☐ Labour saving.

☐ Cost and space effective.

☐ In cold or exposed areas they provide more warmth and comfort for the horse.

☐ Less likelihood of water pipes freezing.

☐ Horses are happier when able to see one another: i.e. herd instinct.

☐ Horses, feed shed, hay store and tack rooms can be under one roof.

☐ Horses are more easily observed should anything be wrong.

☐ The work of student staff is more easily supervised.

Disadvantages and Precautions
FIRE. This spreads more quickly and there will be greater problems from smoke. It is more difficult to get horses out of a barn-type building than out of individual loose boxes. To reduce the risk of horses being trapped, double doors should be placed at both ends of the building. These can be either sliding or hanging doors, and if the latter they should open outwards. The doorways must never be obstructed. In a very long building, doors should be placed half-way down each side to make it easier to lead out horses. Fire drill should be held regularly – see *Prevention and Control of Fire*, page 83. A hay store inside the building, although convenient, is an added fire and dust risk.

INFECTION AND CONTAGION. These are always a problem where large numbers of horses are stabled together and there is a greater risk of this in barn stabling than in a yard of loose boxes. A design which gives good ventilation, large partitions and high roof, reduces the risk. Barn stabling with low ceilings, a warm damp atmosphere and insufficient ventilation is more vulnerable to the spread of disease.

VENTILATION. In all weathers and at all times there must be a sufficient supply of fresh air. The requirement for internal stabling is 56m³ (2000ft³) of air space per horse.

A high roof, as found in the agricultural barn-type of umbrella building, gives ample air space and ventilation per horse and keeps the stables cool in hot weather. If a ceiling is installed, then the diminished air space necessitates the use of some form of mechanical ventilation. Double doors give a through-draught. Windows can be a source of fresh air, but these should be set high in the outside wall of each box and should open inwards, hinged at the bottom. If the building is roofed with asbestos sheets, these can contain continuous roof vents and ridge vents where the roof meets the supporting wall, and louvre boards can be installed in the outer walls to assist the extraction of used air.

COUGH ALLERGIES. These can be a problem if some horses are bedded on straw, and other susceptible horses are on shavings or paper in the same building. It is impossible to keep a spore- and dust-free atmosphere under these conditions, and the susceptible horses are bound to be affected.

BOREDOM. This is rarely a problem as horses can see each other and watch work going on, but, if finance allows, windows or doors 90cms (3ft) square can be put in the back wall at a height of 1.5m (5ft), which when opened allow the horses to look out. They may be constructed of wood or unbreakable clear plastic and covered with a grid. When closed they must be draught-proof.

Conversion of an Existing Building
Most farm buildings can be converted into loose boxes or stalls, provided that the building is sound and the roof is of sufficient height. Low roofs are difficult and expensive to alter and roof height is important, both for ventilation and clearance for the horse's head.

Few buildings provide sufficient space to incorporate

feed and tack rooms. The ventilation may be more difficult to organise, as farm roofs are often tiled or slated. The other advantages and disadvantages are similar to American barn stabling.

Stalls
Stalls are individually partitioned stables in which the horse is permanently tied up, traditionally facing a blank wall. Water, feed and hay are placed in front of the horse.

Some horses do feel restricted and will not settle in stalls, but the majority are reasonably content. Although their movement is very limited, they can still lie down and get up without trouble.

The major advantage of stalls is that they allow more horses to be housed in a smaller space. They are warm, practical, save labour and bedding and are the cheapest type of stable to build. They are specially suitable as day standings for horses and ponies brought in from the field and can be very useful in riding schools.

The conventional design is one line of stalls positioned along the outside wall, with a passageway behind leading to either one or two outside doors. Given sufficient space, two parallel lines of stalls can be built and separated by a 3.7m (12ft) passageway. The more modern design for stalls is two parallel lines facing each other across a feeding passageway. Horses are tied up to the front of the stall, but can see each other through the rails. They are watered and fed from the central passageway. Two wider passageways running behind the horses give room for mucking out and bedding down.

SIZE OF STABLES

Considerations:

☐ Cost – the larger the box the higher the cost for ground, structure and bedding.

☐ Comfort of the horse who must have room to lie down.

Design of Stables

- Ease of management.

- Larger stables can be used for smaller animals, but the reverse is not possible.

- Within limits, the larger the box, the more comfortable for both horse and groom.

- Loose boxes are preferably square, but if rectangular the longer distance should be from door to back wall.

Dimensions

- 3.7m x 3.1m (12ft x 10ft) is adequate for most horses of 16 hands and under.

- 3.7m x 3.7m (12ft x 12ft) or 4.2m x 3.7m (14ft x 12ft) is a comfortable and suitable size for a horse over 16 hands.

- 3.1m x 3.1m (10ft x 10ft) is adequate if under 15 hands.

- 2.4m x 2.4m (8ft x 8ft) for small ponies.

- 4.6m x 4.6m (15ft x 15ft) is needed for foaling boxes.

- 1.8m x 3.3m (6ft x 11ft) for stalls for animals under 16 hands.

- 1.3m x 2.4m (4½ft x 8ft) for stalls if under 14.2 hands.

Roof Height
The roof must be high enough to provide sufficient air space, and to ensure that there is no danger of a horse hitting his head. Horses in outside loose boxes require 42 cubic metres (1600 cu ft), in barn stabling 56 cubic metres (2000 cu ft) and in stalls 28 cubic metres (1000 cu ft) each.

MATERIALS

Materials used in construction must be strong enough to withstand kicking or leaning against walls or attempting to chew accessible wood. Although ponies can be housed in a more lightly made building, it rarely lasts as long, and can be a false economy.

Types and their Advantages
Bricks
☐ Pleasing appearance.
☐ Warmth.
☐ Longevity.
Using bricks is very expensive because, for a strong and weather-proof surface, the walls must be cavity built, i.e. double walls with a cavity between.

Concrete Solid Blocks
☐ Acceptable appearance.
☐ Hard wearing and long lasting.
☐ Minimal maintenance.
☐ Comparatively quick and easy to build.
☐ Cheaper than bricks.

Concrete Cavity Blocks
The advantages of these are the same as those for concrete solid blocks with the additional one of good insulation without extra expense. Although not as strong as solid concrete blocks, cavity blocks are of an acceptable strength for most stables. They can be reinforced with metal strips if they are likely to be subjected to unusual stress.

Concrete Breeze Blocks
These are of an acceptable appearance, are fairly hard wearing and long lasting, comparatively quick and easy to build and relatively cheap. These blocks are porous and only suitable for inner lining. For exterior use they need weather proofing. They are not as strong as solid concrete or cavity blocks, and can fracture under stress.

Prefabricated Sectional Stables of Wood
☐ Pleasing appearance if well maintained.
☐ Readily available.
☐ Easily erected.
☐ Can be moved.
☐ Warm.

There is, however, an added fire risk. Also, wooden, prefabricated sections are more difficult to disinfect, the upkeep costs are higher and they have a comparatively short life compared to bricks or blocks.

Cedar wood is the most expensive material, but lasts longer. A foundation of two rows of bricks to which the structure is bolted is required. All wood should be pressure treated with a wood preservative. Boxes should be lined with suitable hard wood or chipboard.

Although good-quality wooden boxes last 30 years and more, ones of inferior quality show wear very quickly. They can be damaged by horses kicking or chewing the wood, and the resulting cracked or splintered boards can be dangerous.

When selecting wooden stabling, it is advisable to buy the best that can be afforded, and even then to check on the quality of the timber used and the method of construction.

Prefabricated Sectional Stables of Concrete
In recent years there has been some use of this material.

Prefabricated Sectional Stables of Composition Materials
Although not yet used as much as wood, compositional materials may well grow in popularity as their design and structure become better known and proven. They are likely to be cheaper than wood.

Solid Wood
When the log cabin method of construction is used, it is attractive to look at, strong, durable, but very expensive.

INTERNAL WALLS

Barns
The external structure of this type of stabling is a steel- or concrete-framed barn with a clear span of asbestos sheets. External walls can be made of asbestos, plastic sheeting or concrete blocks. The internal partitions making up the boxes or pens can be constructed of brick, blocks, wood,

or prefabricated sectional walls. The last are purpose-built, heavy wooden partitions topped with grilles or heavy galvanised mesh. The mesh is safer as there is no chance of a horse catching his foot when rolling and kicking. These partitions are usually 2.3m (7½ft) in height. Solid partitions between boxes can reduce the risk of infection and contagion. They will reduce draught. The choice of internal partitions is a question of personal preference and cost. The range is from basic but sufficient, to very elaborate designs.

Separate Boxes

These are usually of wood and may go up to the roof or to the eaves. The forweb will prevent dust and spores moving from one box to the next.

LIGHT

Corrugated perspex roof panels can be placed in the roof above each loose box. They will give ample daylight. In hot sun the panels on the side of the roof exposed to the sun will increase the temperature in the stable. Those on a north-facing roof cause no problem. More light can be provided with translucent panels in the outside walls, under the eaves, providing the walls are high enough for the horses not to reach them.

DRAINAGE SYSTEMS

To install a drainage system, planning permission is required, and must accord with local bye-laws. Professional advice should be taken.

The main systems are as follows:

Main Drainage

If this is available, it can be used only for lavatory and tack-room drains. Other arrangements will have to be made for stable drains, roof and surface water.

Septic Tank
This consists of a system of tanks and can be used for the whole of the establishment's drainage. Bacteria present in the first tank work on the material entering it. The liquid then moves on and is eventually piped out under the surface, by which time it is clean. On occasion it may be necessary to empty this system, by arrangement with a local firm specialising in drain cleaning. Some systems require emptying every six months, while others will function efficiently for many years.

Cess Pit
This is a system of one or more tanks which have to be emptied when they are full. A regular contract should be arranged for this.

Soak-away
If already installed, this system will probably be maintained as long as neighbours do not complain. Planning permission is unlikely to be granted for a new installation. The principle is a tank system with bacteria present which work on the solids. The liquid is allowed to drain away across the field by means of a ditch running downhill away from the building. This system works well in a chalk or gravel area, but not in a clay area. Lavatory and tack-room drains must enter the drainage system via a system of sealed pipes.

Traps and Gullies
The principle on which gully traps are designed is for water to separate the outside air from the sewer or main drain air, and so prevent the return of the latter.

Stable yard and surface drains should have efficient traps between the drainage system and the stable yard, so that solids can be collected and removed. There must be a sufficient body of water between the main drain and the air to prevent sewer gas from emerging.

Drains should be well ventilated on the sewer side with a

long pipe going well above the building. This should act as a double insurance against unpleasant smells.

Covered Drains
If possible, these should be avoided, as they are not practical, becoming easily obstructed with manure and straw. If already installed, they should have accessible drain traps, which should be fitted with a removable wire tray to catch any solid material. Alternatively, small-gauge wire netting can be used. Twisted straw can be used for small traps, but must be replaced daily. It is essential that any lengths of covered drain can be easily cleaned with rods. If that is not possible, it should be easy to take up the top covering, i.e. in old-fashioned stalls. A covered drain often runs down the length of the building and must be cleaned out each day.

Open Drains
Loose boxes and barn-type stables can be drained either to the front or to the back of the box. Drainage to the back results in less air contamination. This is particularly important in barns.

DRAINS

In Loose Boxes and Barns
These should have a shallow slope of one in 60, towards either the front or the back of the box. A small opening at floor level in the centre of the wall can direct any liquid into a shallow open drain running along the outside of the stable wall. Traps for collecting straw and solids should be positioned so that no clogging material can reach the main drainage system. The removable trays in the traps should be regularly cleaned. To avoid any draughts from the drain opening, straw should be bedded up the wall.
   The advantages of drains at the back are:
☐ No smell of urine, so more opportunity for the horse to breathe fresh, untainted air.
☐ Drainage is out of sight but still easily checked.

☐ Less chance of draughts, as bedding is always banked up.

The disadvantages are:

☐ Difficult to arrange if loose boxes are on a boundary line or attached to other buildings.
☐ If not easily accessible they are likely to be neglected.
☐ Less easy to wash box down and sweep out water because of smaller opening and no doorway.

Barn-type Stabling

There should be a shallow open drain, preferably along the outside wall to avoid contamination, but if this is not possible, along the inner walls of both lines of boxes. This should end in a gulley to trap straw and manure.

All drains should be washed clean daily. In a very long building, drains should be of the hog-back type, sloping from the centre towards both ends.

Stalls

Stalls should have a shallow open drain 60cms (2ft) behind the back post of the stall.

Deep Litter

Horses bedded on deep litter or any form of solid bed should have any covered drains sealed off. Drainage is not required if this type of bedding is used, but it is convenient to have an outside shallow drain to take away the liquid after washing the floor and walls.

Yard

This should have a gentle slope (according to the natural lie of the land) towards one or more mud gullies or traps which allow surface water to join the drainage system. All drain covers should be of steel or reinforced concrete, capable where necessary of taking the weight of a heavy lorry.

GUTTERING

All stable buildings must have efficient guttering which

is well maintained and regularly cleaned. The down pipes should be of a sufficient number; and shaped to shoot the water away from the base of the building, to open or covered drains. Broken and leaking guttering gives a bad impression of a stable yard, makes buildings damp and shortens the life of any wooden structure.

FLOORS

For the comfort of the horse, and to save strain on the legs, stable floors should be as level as possible. To assist drainage, a slope of one in 60 is suitable, except for stalls, which should be slightly steeper. The necessary floor space is 11m² to 14m² (118 to 150 sq ft).

Materials

CONCRETE is the most widely used surface and is very satisfactory. Minimum depth of 15cms (6ins) is placed on top of a well-drained prepared base of hard core and shingle. A damp course must be included. It is easily put down, but it must be of an extra-hard mixture to withstand urine and the horse's weight, and 'roughed' to provide a non-slip surface.

GRANITE TOPPING makes for a long-lasting surface, but can be slippery. It is doubtful if a grooved surface drains any more satisfactorily than a smooth surface, and it is more difficult to keep clean.

CHALK AND CLAY FLOORS are non-slippery and quiet. They are cheap to put down if material is available locally, but they require more maintenance, having to be topped and rammed every year. A clay floor tends to be damp.

COMPOSITION FLOORING is very satisfactory, but very expensive.

REVOLVING RUBBER MATTING on concrete is used in the USA, and saves time mucking out. The comfort of the horse is not considered. When the horse is out of the box, the flooring is revolved, manure falls off and the floor is washed clean. No bedding is used.

POROUS FLOORS need a well-drained base. The surface is dug out and filled with rounded pebble. There can be a smell problem.

SLATTED FLOORS are under investigation.

HEATED FLOORS are under investigation.

WALLS

Height
RIDGE ROOFS. The walls of most prefabricated loose boxes range in height from 2.1 to 2.3m (7 to 7½ft) at the eaves to 3.3m (11ft) at the ridge. This is adequate for a box with a ridge roof and which slopes up from both front and back walls to a ridge.

When building with brick or concrete blocks, it is advisable for the walls to be 2.4 to 2.7m (8 to 9ft). This will allow for a 2.4m (8ft) doorway.

PITCH ROOFS. A roof sloping from the back down to the front is known as a lean-to or pitch roof, and stables with such roofs require a minimum wall height of 2.7m (9ft) at the front, and 3.3m (11ft) at the rear. In areas likely to have heavy snow, a steeper pitch is recommended. The roof can also be sloped from the front of the box to the rear, when the height of the back wall should be a minimum of 2.7m (9ft).

FLAT ROOFS. Stables with a flat roof or a ceiling with store rooms or lofts above require walls at a minimum height of 3m (10ft), but the measurement is dependent on the ground area of the box. See *Ventilation*, page 28.

Damp Course
All walls should have a damp course; tarred building paper or a strip of plastic paper is effective. This is laid on the foundations, at the base of prefabricated stabling, or a little above ground level with bricks or concrete blocks.

Insulation

The insulation of stable walls is not essential, but adds greatly to the warmth and comfort of the horse. It will also increase the ventilation rate. It is recommended that all prefabricated stabling should be fully lined, to increase the stability, strength and warmth. Effective wall insulation is then achieved by placing a sheet of foil-lined building paper between the outer cladding and the inner lining.

Cavity-wall blocks provide their own wall insulation. Brick buildings constructed with cavity walls, i.e. two layers thick, have full insulation.

A single layer of brick needs lining with wood, and reinforcing with supports on the outside every 1.2m (4ft).

Surfacing of Walls

WOOD STABLING. To give sufficient strength and added warmth, all walls should be lined. This is essential up to 1.5m (5ft) high because of the risk of kicking, and it is recommended that it is continued up to the eaves. Material should be exterior plywood, or a wood of similar strength. Galvanised iron has been used with some success, but it is essential that it is fitted so that no injury from it is possible.

Both exterior and interior surfaces of the wood should be treated with wood preservative. Regular painting of the outside with creosote keeps it in good order. If interior woodwork is painted with colourless preservative, the top half of the box can then be painted with white emulsion. For a shavings or sawdust deep litter, the bottom 60cms (2ft) of the wall, if of galvanised metal, should be painted with black bitumastic paint; if of wood, it should be treated with tar.

PRECAST CONCRETE OR PREFABRICATED COMPOSITION. Such stabling requires lining with wood or chipboard. The appearance of the exterior surface is improved by a coat of Snowcem or similar preparation and the wood treated as above.

SOLID CONCRETE BLOCKS AND CAVITY BLOCKS. Exterior walls require weather-proofing with a proprietary plastic preparation.

BRICK. Cavity walls of brick-built buildings are weather-proof.

The interior walls of both types of blocks and also brick walls are rough. If a horse gets cast, it may damage itself on the rough surface. The walls may be plastered and painted. It should be remembered that lead-based paints are poisonous and should not be used. Plastic paints are now available which can provide a smooth washable surface.

ANTI-CAST RIDGE. This is a ridge standing out approximately 25mm (1in) from the wall surface and 1.5m (5ft) above the floor. For ponies, the height should be 1.1m (3½ft). The ridge enables a horse to right itself, should it get cast, by stopping the feet slipping up the wall. Wooden stabling can include a reinforced strip of wood fixed firmly to the wall with screws. Block or brick buildings can incorporate a row of blocks or bricks set out from the wall, or a metal strip set into the cement between the blocks and bricks. Whatever is used, it should be shaped so that a crib biter cannot seize it with his teeth. Alternatively, a groove can be made in the wall surface to give a foothold to a cast horse.

ROOFS

A roof should meet the following requirements. It must be:
☐ Weatherproof.

☐ Durable.

☐ Noiseless.

☐ Non-flammable.

☐ Warm in winter, cool in summer.

Types
ASBESTOS SHEETING fulfils all the above requirements and is inexpensive. In winter it may need lining to prevent condensation, and for this, roof felting is suitable. Maintenance is nil, unless cracked by a falling bough or heavy blow, e.g. from the horse's head. The dangers of asbestos dust only occur when sheets are sawn, and should be guarded against at this stage. Perspex roof panels can be placed on the north side to provide more light.

RUBEROID, GLASS FIBRE OR PLASTIC TILES AND SLATES have a pleasing appearance, are long lasting, light in weight, suitable for wooden stabling, and inexpensive compared to conventional tiles and slates.

ROOFING FELT is the cheapest material available, but it does not last, and should only be considered if it is of high quality. Wooden strips should be fitted at frequent intervals to prevent it being torn by strong winds.

TILES AND SLATES are satisfactory, but expensive and heavy. They are not suitable for wooden buildings.

THATCH is not recommended because of the fire risk.

GALVANISED SHEETING is hot in summer and noisy in wet weather. It is satisfactory for day stalls if correctly ventilated. Although cheap to install, maintenance can make it as expensive as other roofing materials, as it requires regular tarring or painting to control rust.

RIGID PLASTIC ROOFING SHEETS are light to handle, reasonably long lasting, require little maintenance, and compare favourably in price. They are noisy in rain and high winds. They become brittle in time and are then likely to split.

Overhang
This is a short extension of the roof over the door and front wall area of outside loose boxes. It gives protection to both horse and groom, and is a great boon with no drawbacks.
  With prefabricated stabling, the overhang is usually an

27

extension of the roof and built of the same material. With brick or block stabling, its material depends on construction. It can be built as a continuation of the roof, or else as a separate structure. Translucent plastic sheets can be used, but these can raise the temperature in the box in hot, sunny weather.

CEILINGS

These can be installed:

☐ If storage space or living accommodation is required above stables.

☐ In a high building, to decrease air space and make stables less draughty.

The material used should be fire resistant and, to avoid condensation, be capable of absorbing moisture.

VENTILATION

To keep a stabled horse happy requires a constant supply of fresh air, but without draughts. Except in very cold weather, the temperature in a stable should not be warmer than outside. Horses keep fit and healthy if well rugged. It is preferable to keep a horse warm with extra food, blankets and stable bandages rather than by restricting his supply of fresh air. At the same time, he must not be subjected to a draught, as this can easily give him a chill.

Ventilation is of prime importance when stables are being built or converted. Any deficiencies will adversely affect the health of the horses. Most prefabricated boxes and purpose-built modern barns are not designed with sufficient ventilation. As long as draughts are avoided, the supply of fresh air should never be restricted.

Ideally horses should have a permanent opening of 0.3 sq m (3 sq ft) (inlets) per horse in the walls and 0.15 sq m (1.6 sq ft) in the roof (outlets).

The acceptable air space per horse is 45.3 cu m (1600 cu

ft), and this must be borne in mind when deciding on area and height.

Means of providing ventilation are as follows.

Open Top Doors

These are the main source of fresh air for individual outside loose boxes. Except in exceptional circumstances, i.e. driving snow or a very cold direct wind, they should be kept open. The lower door should be 1.4m (4½ft) to 1.5m (5ft) high, to give sufficient protection and to discourage a horse from jumping over it.

If stables are in lines, then they are best positioned facing south. If the stables have to be built facing north or east, it may be necessary to shut top doors at times in winter. If so, other methods of effective ventilation must be used, such as opening the windows or fixing a short metal hook externally to keep the door just ajar. The hook must be short enough for the horse to be unable to insert his head between the door and the door post.

Windows

Windows are necessary for light and ventilation. If two are installed, they should be placed in the front and rear walls. In very windy weather, if the top door has to be shut, the window at the rear can be left open. In very hot weather, a back window allows a through-draught of air. If only one window is to be installed, it is probably better to have it at the rear.

Windows should be placed as high as possible in the wall, hinged at the bottom or centre according to size, and preferably opening inwards. Cold air is then directed up and above the horse's body, before descending and mixing with the warmer air below. Windows should be fitted with unbreakable clear perspex and protected with a grid which opens with the window. Bars are not suitable as these can be dangerous, easily trapping a horse's foot should he kick up or rear. Glass, if already installed, should be replaced with perspex. Windows fixed at too low a level and opening inwards can be a hazard for the horse. If opening outwards, they must, when fully open, lie flush with the wall. They must be protected by a metal grid so that the horse cannot put his head through.

Louvre Boards

These are broad overlapping boards, set at an angle to prevent the entry of rain or snow, and with an air passage between. They should be fixed in outside walls at least 2m (6½ft) from the ground. They extract stale air, and if correctly fitted, will assist in preventing condensation on the roof and walls.

Eave Vents and Ridge Vents

These are found in asbestos-sheet roofs, at the eaves where the sheets meet the walls, and at the ridge of the roof where the sheets are capped. They are efficient ventilators. They should also be installed in other types of roofs.

Ventilating Cowls

These are placed on the ridge of the roof. They are manufactured in several designs, and are used for roofing other than asbestos. They allow exit of stale, warm air and ingress of fresh air. It is usual to install one to a box or one to two boxes.

Air Bricks

Air bricks can be put just under the eaves in brick or concrete block stabling. They should not be placed lower down because of draughts.

Power-driven Air Extractors

These can be fitted on the roof and should be turned on when the air is very still and humid.

When buying prefabricated stabling, it is advisable to check on ventilation. Some firms make better provision for it than others.

Summary
It is reasonably easy to provide sufficient ventilation in individual boxes but much more difficult in barn stables.

DOORS

There are two types: conventional hinged doors and sliding doors.

Size of Doorway
The ideal size is 2.4m (8ft) high and a minimum of 1.2m (4ft) wide, although for mares, foals and high-class young stock, doors should be 1.4m (4½ft) wide. As an extra precaution against horses knocking their hips, rollers can be fitted on the inside of both door posts. The distance between the posts should then be 1.4 to 1.5m (4½ to 5ft). Many pre-fabricated stables have doors 2.1 to 2.3m (7 to 7½ft) high; 2.3m (7½ft) is acceptable, but 2.1m (7ft) is only suitable for smaller horses.

Conventional Hinged Doors
This is the usual type of door for outside loose boxes. They are divided into two, the bottom door being 1.4 to 1.5m (4½ to 5ft) in height.

FITTING. They should be positioned to the side of the box to reduce direct draughts. They should not adjoin the box next door, as this encourages biting and bullying. They should open outwards so that should a horse get cast against the door it is possible to open it. Outward opening doors also reduce disturbance of the bed. The bottom door should fit firmly down to the floor, so that there are no draughts, and no danger of a horse getting his foot caught under the bottom of the door.

The top edge of the bottom door should be protected from chewing by the horse with a firmly fixed heavy metal strip, which is 8 or 10cms (3 or 4ins) deep on both sides. If there is no overhang, then some protection is needed from the weather for the tops of both doors.

31

Sliding Doors

Sliding doors suspended from an overhead gearing are usually made the same height as the partitions of the boxes or pens of interior stabling, and consist of solid wood to a height of 1.5m (5ft) with grilles above. A sliding door can also be used for outside loose boxes. The horse can look out, but cannot put his head out. A light top door is sometimes fitted to seal the top half in case of need.

Sliding doors cause less obstruction than hinged doors, and are a great asset for interior stabling where horses and people are moving and standing in the central passageway.

FITTING. Sliding doors are heavier than a hinged door, and require substantial posts, usually made of metal. They hang from an overhead gearing, and the door is guided at the bottom either in a groove or on a rail. As a groove easily becomes clogged with dirt, a round rail on which the door glides along a groove in the bottom of its lower edge is more efficient.

Doors at the ends of barn stables, because of their size (approximately 3m or 10ft wide and 3m or 10ft high) must be very heavy, so sliding doors are easier to handle than hinged doors.

Construction

Stable doors receive constant and often rough usage. To withstand this it is essential that they are constructed of strong material. They should be lined with exterior plywood, galvanised sheet metal, or a similar hard, smooth surface. This should also cover the 'stays' on the door which easily damage a horse's leg or knee if left exposed. Inferior wood, particularly if left unlined, is unsuitable and will not last.

☐ Hinges and bolts should be of heavy-duty galvanised iron. Door hinges should be capped to prevent the door being lifted off, and they must be long enough to support the weight of the door.

☐ The bottom door requires a top bolt, and a lower

fastening which can be of the kick-over type. Bolts should be of a simple design, and free from anything which could injure the horse or catch on a headcollar. The ideal door fastenings are those built into the framework, as the horse cannot get caught on them or touch them with his teeth. There are also special horse-proof bolts on the market, which are excellent but expensive. Some horses are able to open doors, so if the bolt is not of the horse-proof type, it should be able to take a spring hook for extra security. The top door requires one bolt. Both doors should have the necessary catches to fasten them back, flush against the stable wall securely.

☐ Because of the danger of fire, stable doors should never be locked.

☐ Window latches must be out of reach of the horse and, if at a lower level, must be so protected that the horse cannot play with them.

STABLE FITTINGS

For safety reasons, stable fittings should be minimal, and in many yards, fixed-feed and hay mangers, hay racks and automatic watering are not installed.

Tying Up

For tying up the ring should be placed to the side of the back wall at a height of 1.5m (5ft) (lower for ponies) and adjacent to the haynet ring, which should be nearer the corner at a minimum height of 1.8m (6ft). A short piece of string may be looped through the tying-up ring, so the horse is tied to the string and not directly to the ring. The loop should be small and neat to discourage the horse from chewing it. A second tie-up ring in the front by the manger can be useful.

A tie-up ring should also be fixed outside the box so that the horse can be secured during mucking out and grooming. Care should be taken that adjoining horses are not tied up at the same time, as they may well kick at each other.

Design of Stables

Rings should be of heavy galvanised steel. In brick or block stabling, they should be bolted through the wall with a holding plate on both surfaces. In wooden stables, the ring should be on a metal plate secured by four heavy screws and fastened through on to the vertical cladding of the box.

Watering

There are a number of ways in which water can be supplied:

• Automatic watering bowls must be securely fixed across a corner and near the door if the drains are at the front. This ensures easy checking, and in the case of an overflow or burst pipe, the bed will not be soaked. All water pipes must be lagged to protect them against frost, and boxed in so they cannot be chewed by the horse. Preferably, they should run along the inside of the stabling above the doors, although if put in during construction of the stabling they can be laid underground.

Automatic watering fixtures can be an added risk if a horse gets cast in the corner or chews and plays with fixtures in the box so they should be boxed in, in a similar way to a fixed manger. Many stables do not use them, particularly those with mares, foals and young stock. Water bowls of the type operated by the horse pressing down a plate with his muzzle are not suitable for horses. Some can be put off by the noise of the bowl refilling and it takes time for the horse to drink his fill.

• Two plastic 14-litre (3-gallon) buckets secured to a ring in the wall by a spring hook or string can be used. The spring hook should be attached to the bucket and not left hanging on the wall.

• Deep, metal containers holding 18.2 litres (4 gallons) can be placed on the floor in a corner of the box with the bedding firmly wedged against them. If they have handles the horse can play with them or catch a foot, and they can bruise a horse's leg should he knock against them. They are awkward to move and being so heavy are unlikely to be emptied and cleaned as often as they should be.

• Plastic buckets with handles are steady and safe for sensible quiet horses if placed on the floor in a corner of the box and firmly wedged with bedding.

• Bucket holders fixed on the wall or at ground level are not recommended, as a horse can easily get a leg caught or knock his head. They are only suitable for stalls.

• Wooden buckets are rarely seen today. They are expensive, heavy and more difficult to clean.

Feed Troughs or Mangers
There are a number of utensils suitable for short feeds:

FIXED MANGERS. These should be solid, with a smooth surface and must be easy to clean. The lip should be broad enough to prevent a horse seizing it with his teeth in case he starts crib biting. An overhanging inner lip prevents food being spilt, but it is difficult to clean. Deep mangers conserve food but encourage a horse to take too large mouthfuls. Shallow mangers are easy to clean, but a horse can easily waste food by pushing it over the edge.

Mangers can be built at ground level. These are safe and create the natural angle for a horse to eat, but they are likely to get soiled with dung and need constant washing.

Mangers are usually positioned at a height of approximately one metre (3 to 3½ft). To prevent a horse catching his foot underneath or knocking his head as he gets up, they can be boxed in, down to and flush with the floor. Alternatively, they can be fixed across a corner with a sloped but solid boxing in by boards. It is essential that there are no projecting edges or angles on which the horse could damage himself.

Mangers should be fixed on the same wall as the door to reduce the risk of anyone being kicked as they leave the box after giving the feed.

In stalls, fixed mangers for food and hay are usually installed at breast height.

REMOVABLE MANGERS. These can be made of metal or plastic with hooks which fit either over a door or into metal fixtures. They can be fitted on the door or on to the walls of the front corner and can be fixed on to any type of wall. These removable mangers are easy to wash and cannot be knocked over. The problems are that if fixed on the door they are high, and at an unnatural angle for a horse eat. Also, they can be lifted out and thrown round the box. *Note*: Metal fixtures are available which run across a corner to hold the manger. These can be dangerous and are not recommended.

Feed Bowls
These can be:

☐ *Metal* which are heavy, less likely to be knocked over, but can bruise a horse's leg should he strike or paw at them. They are easily cleaned, but are heavy to carry when taking round feed.

☐ *Plastic* which are light and easy to clean and handle. They can get knocked over and thrown around the box, or stamped on by a playful horse. They are easily broken. Lightweight, shallow plastic bowls are comparatively cheap, but do not last, and should be removed when the horse has finished feeding. The heavier, deeper feed bowls without handles are more expensive, but more worthwhile. Feed bowls with handles should not be used as a horse can get a foot trapped through the handle. This is particularly applicable for young or playful horses.

Feed bowls fixed in a rubber tyre are safe, but should be removed from the box after the horse has finished eating. They are unlikely to be knocked over and cannot damage a horse's leg, but they are heavy to move about.

Feeding Hay
Hay can be fed in the following ways:

LOOSE HAY is a safe and natural method of feeding hay

and is labour saving. The disadvantages are that it can be wasteful if the horse is an untidy feeder, and does not clean up. Also, the yard is more likely to be untidy, and this method necessitates tying up the hay for weighing, which is difficult if the hay has to be soaked.

HAYNETS are the most popular way of feeding hay as they are economical, tidy to handle, easy to weigh and easy to damp or soak. The disadvantages are that they are time-consuming to fill, tie up and take down; they can be a danger if not tied up correctly (some horses play with the cord, chew it and get caught up); and they are an expense to buy.

FIXED HAY MANGERS are built up from the floor to a height of about 1 metre (3½ft). They must be boxed in, and wider at the top than the bottom. Although practical, they need regular cleaning, which is not easy, and it is more difficult to weigh loose hay.

FIXED HAY RACKS are made of galvanised steel rods or a lighter weight steel mesh. They can be fixed above head level, but dust and seeds tend to fall into the horse's eyes and eating is at an unnatural angle. These problems can be prevented if the rack is fixed breast high, but a horse can then catch his foot or knock his head or eye in the steel rods or mesh.

OPTIONAL STABLE FITTINGS

Grids and Grilles

These are fixed to the top door of loose boxes. They may be a permanent fixture or removable. They are made of steel rods or heavy mesh, and are made in several patterns:

☐ Covering the entire area of the top door, which will prevent horses biting at people or at other horses passing by, help to prevent banging of the door with the front feet, discourage weaving and prevent crib biting on the door. If there is no overhang, it allows horses to see out and yet keep dry.

☐ A V-shaped grid allows the horse to put his head and neck over the door, but prevents him weaving, unless he is the type who stands back in his box and weaves. It also helps to prevent crib biting. Care should be taken that the opening is large enough for the horse to draw back without catching his head.

Salt-lick Holders
These should be positioned in a corner. The danger is that a horse can knock his head or eye on one. Salt blocks can be used as a safer alternative, and placed in a corner where the horse is fed. Salt causes a wet or damp patch, which eventually rots through an unprotected metal or wood wall surface.

ADDITIONAL UNITS

Utility Box
In many yards it is arranged that one or more boxes are set aside as utility boxes. The number depends on the size of the yard. The box may be used for clipping, shoeing, washing or veterinary work.

Requirements:

☐ High ceiling minimum of 3.6m (12ft), preferably 4.3m (14ft), to avoid the risk of the horse hitting his head if resisting when being clipped.

☐ Non-slip, well-drained floor. Rubber flooring is excellent and gives added protection when using electrical equipment.

☐ Daylight from well-protected windows.

☐ Efficient electric light.

☐ Two power plugs.

☐ Water available near by.

☐ Tie-up rings on each wall.

☐ Haynet ring on one wall next to tie-up ring.
☐ There should be no fixed mangers or other fittings.

If heavy clipping machinery is to be used, an overhead trackway is necessary at a height of 3.2m (10½ft) and running along the back and side walls. It is then out of the way, should a horse rear. A corrugated perspex roof panel can be installed to give added light.

Isolation Box

To be effective, this box must be at least 100 yards (91 metres) away from the main stabling. It should face away from the stables, with the prevailing wind blowing away from the stables.

Requirements:

☐ Size: 4.3m (14ft) x 4.3m (14ft).

☐ Height: average.

☐ Roof: insulated.

☐ Walls: insulated.

☐ Windows: well-fitting and draughtproof.

☐ Doors: well-fitting and draughtproof.

☐ Electrics: 13-amp plug and fitting for heat lamp.

☐ Water: adjacent supply.

☐ Adjacent weatherproof shed to store isolated horse's feed, mucking-out tools, grooming equipment, buckets, rugs, medical equipment, etc.

From a practical point of view, it is doubtful if an isolation box is really either worthwhile or effective, but a box that is quite a way from others is useful for ill horses. For this purpose it needs to be easily accessible from the house.

Forge

Stables with a resident farrier require a forge. This should be sited away from the stables, hay and straw barn, because of fire risk. It must be on a hard surface and convenient to other buildings for supplies of electricity and water. In a large yard it is usually adjacent to the carpenter's shop (shed). Most yards now use travelling farriers, with a specially equipped van containing a portable forge.

Useful Additions

HOOKS. A retractable hook outside each box is useful for hanging a headcollar or bridle. It should not be able to be reached by the horse.

SWINGING BALES or poles are sometimes used to separate horses or ponies in standings where there are no permanent partitions. They should only be used as an emergency measure. They can be a danger to junior staff or children who, if in an adjoining area, can get knocked over by the bale or pole if it is kicked or pushed.

Bales are made of timber planks. They are suspended from a roof fixture by means of ropes or chains, and their height can be adjusted.

HANGING CHAIN. For loose boxes bedded with shavings, sawdust or paper a 90cm to 1.2m (3 to 4ft) length of light chain fastened to the outside wall of the loose box makes a convenient place on which to hang rugs and blankets.

ELECTRICITY

• *A 'trip-switch' system is essential* so that if there is any fault in wiring, fittings or machinery the electric current will immediately be cut off, and not be able to be restored until the fault is located and repaired.

• All electric wiring and fittings should be installed by a qualified electrician, who should also attend to all servicing and repairs.

• It should be remembered that horses are much more susceptible to electricity than are humans. Quite a low voltage can kill them.

• All cabling should run through metal pipes and be fixed so that it cannot be chewed or pulled by a horse. Wherever possible, it should run outside the boxes and enter the stables from high in the roof.

• Switches should be of the watertight pattern and positioned outside the box, where they are out of reach of the horse.

Lights
These can be fluorescent or bulb fixtures. Fluorescent lighting must be fixed high in the roof, out of reach of the horse. In adjoining stables, if walls do not reach the roof, one fixture gives good light to two boxes. Fluorescent light is more expensive to buy and install than bulb fixtures, but is cheaper to use, lasts longer, and gives a better light. In very cold weather it may not be so efficient.

Bulb fixtures can also be placed high in the roof and then need not be protected, but this height makes replacing and also cleaning difficult; encrusted dust affects their efficiency. If placed in storage areas, they can be a fire risk should they come into contact with hay or straw. A much safer but considerably more expensive fitting is the bulkhead type. If these are used, they can be placed at a more accessible height 2.1 to 2.4m (7 to 8ft), which makes cleaning and replacement of bulbs easier.

Power Points
One 13-amp power point should be installed and available for approximately every six boxes. They must be out of reach or protected from the horse. They are used for clipping, grooming and heat lamps. They must never be used without a plug-in circuit breaker.

The utility and isolation boxes will need good lighting and an extra socket for the clipping machine, heat lamp, etc.

Yard Lights
Efficient yard lights are essential if work is to be carried out after dark. Good lighting will not only make work areas

41

safer, but will save time and be cost effective. High-intensity tungsten-halogen lamps or high-pressure sodium lamps give excellent light over a wide area. They are cheap to run, although expensive to install. They are very much more efficient than the traditional type of bulb fixture, and require less maintenance. The latter give limited light and are expensive to use.

Yard lights left on at night can be a deterrent to thieves.

MAINTENANCE

Most manufacturers of prefabricated wooden stabling treat the wood used with a proprietary wood preservative, to make it weatherproof, and to protect it from attack by wood beetle and fungus. The woodwork will require regular treatment to ensure that it stays weatherproof and also to improve appearance. Creosote is a cheap and efficient preservative for other woodwork, and also acts as a good disinfectant and fungus killer.

WOODEN DOORS AND WINDOWS in brick or concrete block stables can either be treated in the same way, or painted. This is a matter of personal choice – painting requires more maintenance and is more expensive. The wood requires a primer, an undercoat and a top coat. Lead-based paint is poisonous and should not be used. Maintenance is important, as cracked and peeling paint is a bad advertisement for any yard, private or commercial.

METAL DOORS AND WINDOWS are now likely to be of a rust-proof metal and need no attention. Older fixtures require regular painting, with an anti-rust undercoat and a suitable top coat.

All door, window and stable fittings should be of galvanised iron or steel. These will not rust.

Efficient maintenance of buildings, equipment, facilities, fencing, yard and drive are an essential part of successful business management. This is why it is always better in every sector to buy the best that can be afforded. This does not mean that money should be wasted on inessential buildings or equipment, or luxurious extras which are not necessary. The successful manager plans ahead, and knows what is essential and what is not.

Repairs to buildings and fencing, if carried out in good time, save further deterioration. Allowing surroundings to remain in bad repair invites accidents to animals and staff.

The impressions gained on entering a stable yard is very important. It is possible to tell immediately whether the yard is efficient. Stable managers must be methodical and observant, to ensure that the necessary repairs and regular maintenance work are carried out.

THE TACK ROOM

The tack is a constant security risk. Insurance companies have individual rules about siting and structure. It is advisable to consult your own company before either building a new tack room or converting a building for use as a tack room.

Burglary Precautions

☐ Site to be in view of living quarters and adjacent to or part of the stable block.

☐ Structure to be of brick or concrete blocks with asbestos or tiled roof.

☐ Windows to be of glass, protected on the inside with steel bars.

☐ Doors to be of solid wood or steel, with a normal lock for use during the day and a yale lock for security at night.

☐ Communicating doors should have bolts on the inside.

☐ Yard lights should be left on at night.

☐ Burglar alarms can be fitted.

☐ Dogs have been proved to be the best deterrent, either

sleeping in the tack room or loose in the yard. If the latter, they should have a well-insulated kennel and the yard must be secure so that they cannot wander.

☐ Tack can now be marked with a registered number. This makes identification easier, and discourages theft. In conjunction with the British Horse Society, one scheme is run by Farm Key of Banbury, Oxfordshire, who supply the marking tools and keep a national register.

☐ Insurance companies always require to see security arrangements, and may give advice on further precautions.

☐ It is worth noting that many burglaries take place during the day.

If the yard is to be left, tack rooms should be locked.

Layout of Tack Room

- Wash room for cleaning tack. Main tack room for putting tack up when clean.
- Combined wash and tack room.
- Two combined wash and tack rooms, one for school tack and one for livery.

Whichever system is chosen, the main requirements are:

☐ Floor: concrete is practical and should slope into one corner to drain. If covered with non-slip composition flooring, it is more comfortable for standing and working.

☐ Lighting: fluorescent strip lighting is the most efficient.

☐ Heating: there must be no direct heat to leather. All heating should be well above any working areas to avoid any risk of fire and injury. Electric wall heaters fixed high on the wall are suitable, and the fan-heater type is efficient and safe. The temperature suitable for working, and to keep tack in good condition, should be between 13 and 16°C (55 and 60°F) in winter.

☐ A sink with extended draining board for scrubbing. A deep sink makes cleaning easier.

☐ Hot and cold water: electric water-heaters of the geyser type are suitable. It can sometimes be arranged for room heating and water heating to be on the same system, and connected to other areas requiring heat and hot water.

☐ Saddle-horses for cleaning saddles should be sufficient in quantity for the number of staff. They can be 3 to 4.2m (10 to 14ft) long and must be strongly built.

☐ Telescopic bridle-cleaning hooks: these should be sited with care, as they can be a hazard if placed near a door or where others are working.

☐ Saddle racks, preferably of plastic or wood, as iron will mark the leather. Slip-in holders for horses' names make for efficiency.

☐ Bridle holders, preferably semicircular in shape so that the top of the bridle is kept rounded. Slip-in holders for the horses' names are advisable.

☐ Hooks for lungeing tack, spare girths, stirrup leathers, martingales, breast plates, etc. Name plates help to avoid disorder.

☐ Bit board for spare bits.

☐ Cupboard for tack-cleaning equipment.

☐ Cupboard with shelves or compartments for bandages, boots, shoeing tools, clipping machines, disinfectant, shampoo, hoof oil, Stockholm tar, etc. The top of the cupboard can be flat to provide a sound working surface.

☐ Medicine cupboard for first aid: thermometer, kaolin poultice, salt, drying-up lotion, Epsom salts, cotton wool, surgical bandage, paraffin gauze, wound powder, wound ointment and methylated spirit.

☐ Drugs, poisons and worm powers must be kept in a locked cupboard in the office.

☐ Cupboard for grooming kit.

☐ Notice board for yard rules, daily ride sheet and work sheet.

☐ Rug and travelling boxes around walls.

☐ Several bar stools can make convenient and space-saving seats.

☐ Lockers for staff. Alternatively, these can be in the staff room.

☐ One or two roller towels.

☐ Battery clock.

☐ Dust bin.

☐ Fire appliance.

N.B. The Health and Safety at Work Act (1974) should be checked.

ADDITIONAL PREMISES

Wash Room
Large yards will benefit from a wash house.

Requirements:

☐ Drying racks.

☐ Washing machine and spin dryer.

☐ Heating to assist drying.

☐ Large sink with hot and cold water, and a draining board.

Store Room
If space is available, a store room is useful for spare rugs, travelling gear, spare tack, replacement goods for the tack room, feed shed and yard, etc.

Staff and Lecture Room

The need for separate staff and lecture rooms depends on the number of staff, and the size and the functions of the yard.

Requirements:

☐ Efficient heating.

☐ Floor covering, which is easily cleaned but provides some warmth.

☐ Sufficient upright chairs for students and working pupils.

☐ One or more tables.

☐ Small easy chairs.

☐ Electric kettle, small serving table, power point.

☐ Blackboard and chalk.

☐ Wall charts to cover points of the horse, skeleton, muscles, main organs of the body, bits, etc.

☐ Specimen shoes.

☐ Specimen bones of the horse's leg below the knee.

☐ Notice board; coming events, competitions, etc.

☐ Battery clock.

☐ Long mirror by the door for use of staff and working pupils.

☐ Fire appliance.

☐ Kitchen and small electric stove and sink with hot and cold water.*

☐ Staff and working pupils' individual shelves or lockers.

☐ Hot drinks machine.*

☐ Video camera and power point.*

* *not essential in a small yard.*

Notes: The facilities must comply with the Health and

Safety at Work Act (1974). The minimum requirements are stated in the Horse Riding Establishment's paper – *Guidance on Promoting Safe Working Conditions* (HM Agricultural Inspectorate Health and Safety Executive 1988).

Changing Rooms and Lavatories

These must be sufficient for the size of the yard.
Flush lavatories must be easily available for clients.
Requirements:

☐ Changing room.

☐ Efficient heating.

☐ Floor covering which is easily cleaned.

☐ Basin with hot and cold water.

☐ Shower room, but this is not necessary in a small yard.

☐ Towels, soap and paper towels.

☐ Hooks and coat hangers for clothes.

☐ Boot jacks.

☐ Mirror.

☐ Equipment for cleaning basin, lavatories and floors.

NB If areas requiring hot and cold water are close together, i.e. feed room, kitchen, changing rooms and lavatories, it is more economical as one water heater can be used.

Tack and Clothes Shop

Many yards run a tack and clothes shop. This can be limited to a few items, or cover a large selection of goods. It can be a useful service for clients, and a source of extra income to the school.

Requirements:

☐ A suitable room, which is dry and essentially burglar proof.

☐ A member of staff with sufficient time to cope with customers.

☐ Arrangements to be made with manufacturers to supply goods at wholesale prices.

☐ A trading licence.

☐ Cupboards, shelves and counter.

☐ Cash box.

☐ Saddle racks.

☐ Bridle and tack hooks.

☐ Boot jack.

☐ Mirror.

☐ Tape measure.

Such a shop may alienate the local saddlers who do the repairs, so it might be advisable to make arrangements for them to supply the goods at a discount rate.

Feed Shed

Requirements:

☐ It is best constructed of brick or concrete blocks to discourage vermin.

☐ A dry, light, secure room, convenient to the stables. A good door is essential, so that there is no danger of a loose horse getting into the feed shed.

☐ Large yards may have either a small feed shed for daily use, and a larger feed store next door for storing bulk supplies, or separate feed sheds for each area.

☐ The floor should be of concrete, level except in the door area, where it should have a slight slope and shallow drain to the side of the door.

☐ Barley boiler and bin for soaking sugar beet which

should be near the door to prevent any seepage of water back to the rest of the food supplies.

Additional Equipment for Big Yards
- Sink with hot and cold water.
- Wooden pallets in which to store food sacks.
- Vermin-proof metal containers to hold 3 to 4 tonnes of grain.
- Electric mill for bruising corn, and face masks for operator if machine is not fitted with dust extractor.

CHAPTER 3
Riding Areas

INDOOR SCHOOLS

With the increase in number of riders, the demand has grown for all-weather teaching areas and competition facilities, especially during the winter months. An indoor school has become a necessity for any progressive commercial establishment.

The construction will be governed by the amount of space and money available, and also the purposes for which the school is required.

Planning permission is necessary. If a large, competition school is planned, understandably, local residents often object to the inevitable increase in horse traffic. The tactful canvassing of local support before applying to the planning authorities can possibly avert the refusal of a planning application.

There are now many well-established and experienced firms who will advise on construction, and also deal with the planning authorities. Before a decision is made, it is advisable to investigate a variety of types of schools. Owners are often happy to show their schools and discuss design and structure.

Size
- 20 metres wide x 20 or 30 metres long.
  This area is suitable for teaching small children, adult

beginners, for lungeing horse and rider and training horses on the lunge. The minimum width should be 20 metres, both from a functional point of view and so that, if at a later date a school of the size of a dressage arena is required (20 x 40 or 20 x 60 m), extra bays can be added. Should lack of space prohibit this width, a school 15 metres wide is still a great asset.

- 20 metres wide x 40 metres long.
  This size of school is suitable for general teaching, schooling of horses and small dressage and show-jumping competitions. It is the most convenient and economical size for practical work. It is also the minimum size acceptable to the British Horse Society if the school wishes to be recognised as an examination centre.

- 30 metres wide x 90 metres long.
  This size of school is suitable for all competition work, and also divides comfortably to accommodate three class lessons. The increase in width adds considerably to the expense, as the supports of a wider clear-span roof have to be so much stronger. There is also the inevitable increase in rateable value. A large school fully enclosed with a seated gallery has a much higher rateable value. A smaller school with semi-enclosed walls, small unfitted gallery and limited competition facilities should warrant a much lower rating.

Construction
Most modern schools are of clear-span, steel-framed construction with a ridge roof of asbestos sheets. Translucent sheets are placed at regular intervals in the roof to provide light. The ridge of the roof is capped, and this should be ventilated. The height to the eaves is 4.3 to 5.5m (14 to 18ft). Fourteen feet may be taken as a safe minimum. When floor material is added, the actual roof height will be less. The outer cladding of the walls of the building can be breeze blocks, galvanised metal sheeting, rigid plastic sheeting, asbestos or brick. Brick is very expensive, but sometimes

insisted upon by planning authorities. If asbestos is used at ground level it may be fractured by a blow. Asbestos has lost popularity due to the health risks involved when working with it.

If the walls are built up to the eaves, translucent sheets should be used on the top section to give light. These can be put on to frames which can be opened in hot weather to give essential ventilation. A fully enclosed school can become airless and an unpleasant place in which to work. A cheaper and more pleasant alternative is to have walls 2.4m (8ft) high with an open space to the eaves. The walls may then be of solid timber, and an outer cladding is not required. This system gives sufficient shelter in most winters, is lighter, and allows a greater feeling of freedom to both horse and rider. In the summer, the school will be cool and pleasant to work in, even in hot weather. Schools used regularly for winter competitions, or situated in heavy-snow areas, may have to be fully enclosed, as do those where it is desirable to shut out external noise.

Kicking boards on the inner side of the wall cladding should be of solid timber (railway sleepers are suitable), with a minimum of 1.5m (5ft) showing above the floor. They can be sloped inwards from the top at an angle of 12 to 15°, which will reduce the risk of a rider knocking a knee. To give added light they should be painted white. Dressage markers on the walls should be painted black.

If funds are limited, an alternative to solid walls is straw bales or woven fencing protected by a guard rail. Care must be taken when erecting these to ensure the rider cannot catch his foot, and a watch should be kept for any loose baling string. If a rider or horse falls against these they provide a more cushioned landing than a solid wall.

School doors should be a minimum of 3.7m (12ft) wide and at least 2.4m (8ft) high to avoid any danger of a horse jumping out. If regular topping up of the surface is envisaged, then doorways must be high enough to take the transporters. They should preferably be of the sliding door design, suspended from an overhead gearing. Hinged doors are heavy to handle and a problem in high winds. If used, they should open outwards, as inwardly opening doors are a danger to horses working in the school.

Drainage
The roof needs efficient guttering and drainage. If other buildings allow, soak-aways can be dug at intervals round the outside of the building to take this roof water. The soak-away should be a minimum of 4.5m (15ft) deep and filled with stone or pebble. Alternatives are a 'French drain' along both walls or a shallow open drain to direct the water away, possibly to the yard drainage system. A 'French drain' starts at a minimum depth, and is dug out and sloped towards a deep soak-away. The drain and soak-away are filled with stone or pebble.

The floor area, unless at a high water-table level, should only require levelling before being covered with the selected surface. It is necessary to take advice, but the extra expense of a prepared base and a plastic membrane is often avoidable. It should be checked that surface water from an upper level cannot seep under the school floor.

Galleries
These can be either at the end or along the side of the arena. For watching and judging jumping a side gallery gives better viewing. For dressage judging an end gallery or box is essential. For competition centres, a tiered gallery is recommended. For smaller centres where the gallery may be used for several purposes, a level floor is of more practical use. If regularly used in winter, some form of heat is needed. Fire regulations must be observed.

Lighting
This can be a very expensive item, and research is advisable as to the cheapest and most efficient system to install, maintain and use. Strip lighting has been the most popular, but improved alternatives have now become available.

Teaching Aids
A full-length mirror or mirrors in the corners and on the centre line and another, wider mirror at the half marker on the longside are excellent instructional aids. A video camera is an expensive but useful asset. A loud-speaker system is of assistance, for competition and instructional sessions where both the spectators in the gallery and the riders need to be able to hear.

Jump Store
It is convenient to have jumps readily available, and essential that they are stored in a secure, dry area. If space is available, the area underneath the gallery can be used, with a door in the school wall. For safety this should open inwards to the store and away from the school track.

Surface
Choice depends on the proposed use of the school and the cost. Before deciding, it is advisable to visit other schools to inspect and ride on the surface they have chosen, to check on maintenance, and to consider whether it will stand up to the type of work required.

SAND AND SHAVINGS make a pleasant surface on which to ride. It is suitable for heavy work and jumping. The sand is laid to a depth of about 10cms (4ins). This is then topped with the same depth of shavings, which are rolled level. In time the two surfaces will mix, and can be topped up as required with either shavings or sand, according to the texture of the floor. Sand provides more substance and a heavier surface on which to ride. Shavings make it lighter.

SHAVINGS are at first laid to a depth of 15cms (6ins) and then rolled to make a solid base. As they settle they should be topped up to a final depth of 30 to 60cms (1 to 2ft), to ensure that, when jumping, there is no danger of getting through to the ground surface. They make a good general-purpose surface at a reasonable cost, particularly if the shavings can be bought or collected locally. But there will be a problem of dust in summer – see *Control of Dust* below.

MANUFACTURED WOOD FIBRE is now supplied by a number of firms specialising in riding surfaces. Many will insist on a plastic membrane, particularly if a prepared base has been put down, as this will prevent stones working up. If the base is topped with very small pebble, this problem should not arise. In time, plastic membranes work up and can wear through, particularly with regular jumping use. They are often an unnecessary expense. The wood fibre is spread, levelled and saturated with water. It is then compressed by a vibratory roller. The resultant surface is pleasant and resilient to ride on, particularly for working on the flat. It is not so suitable for jump training and lungeing, as the surface can be slippery, and will also get cut up and need considerable maintenance.

SYNTHETIC MATERIALS. Many new types of surface are now available. They are more expensive but are harder wearing.

Maintenance of Surfaces
All surfaces, given sufficient use, will require daily levelling. This can be done by hand, or by a tractor pulling a harrow and leveller. Specially designed machines for this are now available. Particular attention should be paid to the outside track, the lungeing area and any part where jumping regularly takes place.

Some surfaces, particularly after several years' use, may become compacted and slippery. They will require scarifying and then levelling, and in damp weather this may have to be done regularly. They should also be topped up with fresh supplies. A scarifier is a deep type of harrow which digs up the under-surface. It should not be used if a membrane has been put down.

Control of Dust
In time, school surfaces break down into smaller particles, with resultant dust. In summer, and also in a dry, windy

spell in winter, control of this dust is a constant problem.

Preventative measures:

☐ Surfaces made from wood will need regular applications of agricultural salt, approximately one tonne per 20 x 40m arena. Salt is usually supplied in sacks of 50kg and can be most easily spread if carted on the back of the tractor and thrown out with a shovel. Care must be taken that not too much salt is applied as it can have a drying effect on horses' feet, with resultant problems. It may also affect the leather boots or shoes of anyone standing in the school for a length of time.

☐ Daily watering by hose pipe, either by hand or by use of a garden sprayer.

☐ Installation of an overhead watering system, which is expensive but efficient, quick and labour saving.

☐ Sump oil spread with a fine spray effectively lays the dust for many months. The surface is oily for a few days, but the oil is then absorbed and appears to cause no further problems. The main difficulty is the spraying of the oil so that it is evenly spread over the surface. This treatment is invaluable in drought years when watering is forbidden.

OUTDOOR SCHOOLS

Many firms now specialise in the layout and construction of outdoor riding surfaces. If requested, they advise and provide a complete package deal, and some give a guarantee. It can be a short-sighted policy to try to cut costs, and finish with a school only usable in dry weather.

Considerations when Choosing Site

DRAINAGE Efficient drainage is essential. A site on a slight slope with natural drainage to a lower level can be a great asset. Sub-soils of chalk, gravel, sand or stone all drain well, which makes the site preparation cheaper. Schools built at the bottom of a slope, or on a sub-soil of clay or other poorly draining material, need considerable work and extensive foundations if the final top surface is to be satisfactory and remain usable in wet weather. Experienced advice should be taken, and no expense spared on this basic preparation of the site.

ACCESS. Preferably this should be convenient to the stable yard, with a hard or firm access to a roadway. A muddy track makes horses, staff and spectators unnecessarily dirty. Mud carried on and in the horses' hooves will not improve the school surface.

SHELTER. In exposed areas, neighbouring buildings and/or barns can give welcome protection from high winds and driving rain, but they should not be so close as to interfere with drainage plans. A tall hedge makes an excellent wind-break. The surrounding fence can be built of wooden boards or plastic strips, and to a sufficient height to give shelter. As long as there are gaps between the boards, it will be unlikely to be blown down.

Trees in summer can give welcome shade, but if large they may interfere with drainage plans.

Fencing
The school must be safely and securely fenced according to its intended use. A suitable height for most purposes is 1.4m (4½ft).

☐ Posts should be on the outside of the fence; the inside should have no protruding surface liable to cause injury. For economy, plain wire can be used below a top rail of wood or plastic. Barbed wire must never be used.

☐ The entrance should be wide enough for a tractor and any maintenance equipment; slip rails or a gate can be used. The gate should open outwards, and have no protruding fastenings.

☐ The school surface will need to be retained by a ground-level edging of concrete blocks or strong timber.

Choice of Surface
This depends on intended use, and the amount of work done on the school. Suppliers of surfaces should be given full information, so that they are in a position to offer the most helpful advice. It is also sensible to see different surfaces under the working conditions similar to those envisaged. Most firms will supply a list of schools they have constructed, and owners are usually happy to discuss matters. It is a help either to ride on the surface, or to see horses working on it. Jump training and lunge work are likely to cause the most problems.

It is essential that the correct type of foundation is laid to suit the top surface. This is one reason to arrange for the work to be done by a specialist firm, who will also have experience of blending different materials to produce surfaces suitable for particular work.

All surfaces must be put down on a properly prepared base, which should be built up to be well above the level of the surrounding ground surface.

SAND. This has been the most widely used all-purpose surface, and probably in the long run remains the most satisfactory. If laid too deep, it can be heavy to ride on and tiring for the horse. A starting depth of 8 to 10cms (3 to 4ins) is recommended, and this can be topped up as required.

In summer, it can be dusty without regular applications of salt and daily watering. In winter, if drainage is faulty, it may freeze, but salt helps to keep the school usable.

The sand used should be clean, and free from clay or other soil which will make it compact under pressure.

Types of sand:
(a) Sharp sand double-washed provides a workable surface.
(b) Silicon sand is the best. It is more expensive, and in some areas may be difficult to obtain. It is clean, dust-free, drains well and is therefore less likely to freeze.
(c) Grade 4 industrial sand is also recommended.

(d) Black sand provides a good surface, but tends to be dusty in dry conditions and unpleasant when wet. It must be thoroughly screened to remove all foreign matter.

Mixtures of sand with shavings and plastic fibre are also workable surfaces, the shavings/fibre making the surface lighter to ride on and less likely to freeze.

SHAVINGS. These are put down on top of a solid base. They are then soaked by hose and compacted with a vibratory roller. The resulting surface is suitable for light work, but lungeing and constant jumping are likely to break it up.

MANUFACTURED WOOD FIBRE AND WOOD CHIPS. Used on their own they are suitable for work on the flat. If mixed with sand, they form an improved all-purpose surface. In freezing conditions, shavings, wood fibre and chips, if well drained, will soften when ridden on and should not freeze.

All forms of wood-based surfaces will eventually break up and start to rot, they will then have to be removed and replaced – a costly operation.

PLASTIC. Various forms of plastic granules are now obtainable. They form a clean, dust-free, long-lasting surface which drains well and is frost resistant. They are not so suitable for jump training.

RUBBER. Shredded rubber products give a resilient non-skid surface requiring little topping up and are frost resistant. The shredded rubber is laid on a base of washed stone aggregate which mixes with it to form the surface.

ASH AND CLINKER. This surface is now only available in certain areas. It is practical and cheap (delivery charge only). However, in time it consolidates into a very hard surface, dusty in dry weather and unpleasantly dirty when wet. It freezes hard in cold weather.

Membranes
A membrane has two distinct uses:

☐ *Non-porous type* This is laid down before the foundation. It prevents water coming up, and on a clay soil stops the foundation and school surface from gradually being absorbed into the clay.

☐ *Porous type* This is laid on top· of the foundation and below the top surface. It allows surface water to drain through and prevents stones from working up. Porous membranes will in time clog and cease to drain.

All membranes will in time disintegrate and their original function cease.

CHAPTER 4

Organising and Running a Yard

THE OFFICE

A well-organised and well-equipped office is essential for the efficient running of any equestrian establishment, whether it is a riding school, livery yard, training and competing yard or racing stable. It is needed to give a good impression to clients, and to facilitate the day-to-day running of the yard.

The suggested equipment is detailed below. Some of this will not necessarily be required for all establishments.

Equipment

Flat-topped desk with drawers and chair.

Some form of heating.

Telephone.

Intercom. system between office, tack room, yard, and indoor school.

Outside telephone bell.

Telephone answering machine.

Typewriter.

Cash box with float for change.

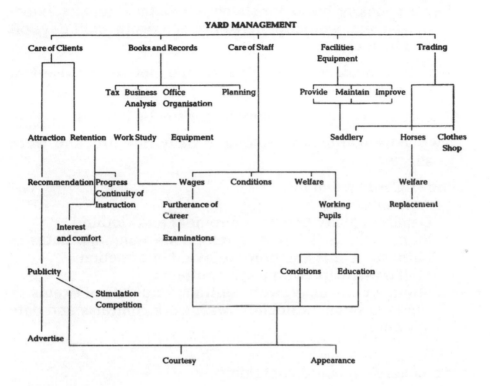

Correspondence tray.

Telephone book.

Tidy tray for pens, pencils, paper clips, etc.

Stapler and hole puncher.

One or two spare chairs.

Wall chart.

First-aid box.

Waste-paper basket.

Desk diary with a day to a page.

Record book or books for shoeing, vet, tack, repairs, injections, teeth, worming, horse hours of work, staff days off and holidays.

Accident/incident book. This should not be a loose-leaf book.

Accident forms for use should a person be injured.

Two loose-leaf books containing daily ride lists and work sheets.

Index Card system for:
Accounts.
Details of livery horses, equipment and clothing.
Riding clients' addresses, telephone numbers, dates of birth, standards and other relevant information.
Staff particulars, with work contracts.
Student particulars, with contracts and date of exams.
Working pupil particulars, with work contracts and date of exams.

Stationery cupboard containing:
Printed letterheads and cards.
Brochures.
Typing paper.
Desk supplies.

Cash books:
Main cash book.
Petty cash book.
Wages book.
VAT book if the business is registered.
PAYE tax tables.

Index file system for invoices, receipts and correspondence.

Notice board with:
Riding school licence.
Third-party insurance certificate.
Business licence.
Employers' liability certificate.

Fire instructions.
Accident instructions.
Telephone numbers of doctor and veterinary surgeon.
Information on school activities.
Information on local activities.
Staff instructor certificates.

In large yards, additional equipment for increased efficiency:
Photocopier.
Tape recorder.
Word processor.
Computer.
Facsimile machine.

ACCOUNTS

Unless the proprietor or secretary has experience of accounting, it is advisable for all businesses to employ an accountant who can:

☐ Present a true record of the business to the Inland Revenue.

☐ Claim all available tax allowances.

☐ Advise as to the present state of the business, future prospects and trends, as well as providing essential information for constructive future planning.

The accountant may be asked to present the books monthly, three monthly, half yearly or annually, according to personal preference and the establishment's ability to maintain good records. As he will be paid by the hour, the more that can be done to keep the accounts in good order, the lower his bill will be. He will advise on the system that he wishes to use.

The accountant will require the following books and information:

Main cash book. Receipts are noted on the left-hand

page, payments on the right-hand page. All are totalled monthly.

Petty cash book for recording small items paid for in cash. The total is transferred to the main cash book each month.

VAT book. For recording inputs and outputs of Value Added Tax.

Wages book.

Bank statements for current and deposit accounts.

Bank cheque stubs.

Details of creditors.

Details of debtors.

Records of livestock: i.e. list of horses, with details of purchases, sales or death.

Records of stock: i.e. machinery, tractors, cars, trailers and horse boxes, with details of purchase, sale and writing off.

Stocks of hard feed, hay, straw and bedding.

INSURANCE

Riding is a hazardous sport, and accidents will occur in spite of every precaution taken by the riding-school proprietors. Insurance policies must therefore be taken out.

Compulsory Insurance

Third-party insurance is compulsory for riding schools, and is recommended for all types of yard.

Employers' liability.

Cars, horse boxes, trailers.

Tractors used on a public highway.

Recommended Insurance

Fire.

Theft of horses, saddlery, machinery, furniture, personal belongings.

Loss of business due to fire or theft.

Personal accident and illness to proprietor and/or chief instructor.

Personal accident insurance for clients.

Office insurance.

NB: Livery horses and their equipment should be insured by their owners. They should not be accepted by the establishment until fully insured.

Non-essential Insurance

SCHOOL HORSES. It is usually considered uneconomic to insure school horses against accident and illness, as premiums are expensive and cannot easily be claimed.

EPIDEMIC. Causing loss of business. It is a matter of personal choice and type of business as to whether this is considered worthwhile. Several insurance companies arrange Umbrella Policies which are worth investigating for riding schools and livery yards. BHS-approved schools receive very favourable terms from the BHS brokers.

CLIENTS' PERSONAL ACCIDENT INSURANCE. In recent years there has been a considerable increase in injury claims by clients riding at riding schools. Many proprietors now recommend that clients should take out some form of personal accident policy before either learning to ride or continuing to ride as a hobby. This does not exonerate the school from legal action should an accident occur and negligence be proven.

PRIVATE HORSE OWNER'S LIABILITY. The private horse owner is advised to insure against third-party risks (at present this is included for BHS members), and they should preferably insure themselves against personal accident.

FREELANCE INSTRUCTORS. Independent freelance instructors are also advised to insure themselves against third-party risks and personal accident.

Incident or Accident Book
It is essential for all types of yard to keep an Incident or Accident Book. This is a compulsory measure for BHS- and ABRS-approved schools. The book must record the time, place and date of any incidents or accidents, an account of what took place, and the name and address of any witnesses. It is often months – even years – before information on an accident is asked for. During this time, important details can be forgotten, staff may leave, and reliable information may thus be difficult to obtain. Suitable accident forms can be obtained from the BHS.

No riding establishment should be left in the charge of a person under sixteen years of age. It is the responsibility of the owner, proprietor or stable manager to decide when a person is sufficiently responsible and capable of being left in charge. The same responsibility applies when allowing riders to be escorted or pupils to be taught by young people without supervision. At present, litigation is all too easy, and court cases and claims for damages are all too common, so it is essential for responsible proprietors or stable managers to take extra care. For example, a notice stating that: 'Although every care is taken, there can be no responsibility for accidents' gives no protection should a claim for accident and injury lead to litigation. The account of an accident should be signed by the person in charge, *and*, whenever possible, by the injured party and any witnesses. For legal requirements, see Chapter 6, page 94.

DRESS FOR RIDING CLIENTS

There should be strict rules as to suitable and safe dress for clients riding at any establishment: though correctly fitting BSI standard hats should be worn, with attached chin-strap securely fastened. Under the Health and Safety at Work Act the wearing of a BSI 6473 or BSI 4472 standard hat is mandatory for all employees and students at Approved riding schools. *The casual loan or hire of hard hats is not recommended.* If, as the result of an accident, the hat should come off or be found faulty, the proprietor of the school could be claimed against for having provided it.

Shoes or boots should be worn with smooth soles and small heels, so that if the rider falls:

☐ The foot will not be caught up in the stirrup by ridges in the sole.

☐ The low heel will prevent the foot from slipping through the iron.

The following footwear should not be allowed:
Any ridged Wellington boots or muckers.
Buckled shoes.
Tennis, walking or game shoes with a solid sole and no heel.
High-heeled boots or shoes.

Floppy plastic raincoats which in a high wind can blow about, making a sudden noise and frightening a horse, should also be forbidden.

ORGANISATION OF FACILITIES

Indoor School

☐ School walls must be well maintained and of a suitable height and strength. No part of the surface should protrude.

☐ School doors should be in good working order.

☐ The gallery must be well fenced off and safe.

☐ Jumping equipment and jump cups should be kept out of the way of riders, or be placed so as not to be a hazard.

Outdoor School

☐ Outdoor schools must be kept well fenced. Post and rails or plain wire with a rail above are suitable. Barbed wire must never be used.

☐ The entrance should have a gate. It should be well maintained and easy to open and shut.

☐ Broken boards or rails should be immediately repaired or made safe, so that no jagged edges, boards or nails are left exposed.

☐ Jumping equipment should be as for indoor schools.

Lavatory Facilities

Suitable facilities must be provided. Wash basins should preferably have hot and cold water. Soap and towels should be available.

Catering

Establishments providing food for staff and/or clients must make sure that their arrangements comply with the Food and Hygiene Act (1900). This includes suitable arrangements and facilities for the storage, cooking and serving of food, and for washing up afterwards. The regulations of the Fire Precautions Act (1974) should also be consulted.

Organisation of Tack Room

Efficient organisation is essential.

• All saddles and bridles should be clearly marked with the horse's name. Metal discs can be clipped to the near-side front 'D's of the saddle, and to the buckle on the throat-lash of bridles. It is then possible to change them if necessary. An alternative is adhesive tape placed under the skirt of the saddle or under the cantle to the rear of the saddle.

• There must be an adequate number of saddle racks; saddles placed one on top of the other can be easily knocked to the ground and may break a tree. Racks for school tack can be arranged along a wall in three or more layers.

- Bridles can be hung separately on another wall.

- Tack-cleaning equipment must be provided for each member of staff involved in cleaning. Time is wasted if supplies are short.

- There must be facilities for the drying of numnahs, rugs, staff coats, etc. Old-fashioned drying racks, worked on a pulley system and suspended from the ceiling, are space-saving and efficient, but they must not be sited too close to any upper wall-heaters because of the fire risk.

Livery Tack
Each owner's saddles, bridles, martingales, etc. should be kept together, with the individual grooming kit below them, so that there is no muddle on a groom's day off. Livery owners should keep their spare tack and rugs at home, unless space is available for their own rug box. All livery tack and belongings should be clearly marked by the owner.

Organisation of Feed Room
☐ Feed sacks, if stored on pallets, should be positioned so that there is space for cats to move round all sides and underneath.

☐ Floors should be swept daily and kept clean.

☐ Sink and feed bowls should be cleaned daily.

☐ Fresh supplies coming in should be placed behind the sacks, so that the old food is used first.

☐ Rats and mice can be a serious problem in feed rooms. Apart from the problem of loss of feed, the droppings of rats and mice in the feed eaten by a horse can cause severe colic. Stable managers must be aware of the problem and take suitable action. Some of the poisons used in the past are no longer effective, and farm cats can be far more efficient.

FEES

Fees for Rides and Lessons
The amount charged should depend upon:
(a) Keep of horse—feed, bedding, water, electricity.
(b) Labour—mucking out, grooming, tack cleaning.
(c) Instructor or escort time.
(d) Shoeing and veterinary expenses.
(e) Saddlery—upkeep and replacement.
(f) Horse replacement.
(g) Share of overheads—rent, rates, bank interest.
(h) Share of upkeep of box, yard, fields, fencing, water, cultivation.
(i) Share of insurance and office expenses.
(j) Local conditions.

Livery Fees
Items to be taken into account:
☐ Items (a), (b), (g), (i) and (j) above.
☐ Exercising time if required.
☐ Profit should be a minimum of 10 per cent. Livery-only yards must take a bigger percentage if it is to be their main source of income.

Schooling Fees
These depend on the riding time of skilled staff.

Working Liveries
The advantage of these is the use of an extra horse without additional capital expense, but before agreeing to this system it is essential to ensure that the horse will be really useful. Remember, too, that the box taken up by a horse could have been occupied by a full livery.

Clear arrangements must be made with the owner as to:

☐ Days and time when the horse will be available for use in the school,
☐ Responsibility for clipping, shoeing, veterinary expenses, tack repairs.
☐ Responsibility for livery if the horse is ill or off work.

Fees should depend on individual arrangements, but are usually half to three-quarters of the full livery fee. The livery owner's insurance company should be notified of the arrangement, as this may affect the terms of the policy.

Do-it-yourself Livery
This is an acceptable arrangement for the owner of the yard, and the owner of a horse, providing that arrangements are clearly defined between both parties.

STAFF

Staff Contracts
Within thirteen weeks of starting work, all staff should have a work contract. This should include:

Job description.
Date of commencement.
Hours of work – but as the care of animals is involved these have to be flexible.
Days off per week.
Arrangements for holidays, including statutory holidays.
Pay per week or month.
Overtime.
Sickness arrangements.
Arrangements for termination of employment on either side, and disciplinary and grievance procedures.
Special arrangements: i.e. board and lodging, living-out allowance, keep of own horse, use of car or Land-Rover and trailer, time off to compete, preparation for exams and keeping of pets.

Employers are reminded that arrangements for terminating employment due to unsatisfactory work or behaviour should fulfil the guidelines laid down in the relevant pamphlets published by the Department of Employment.

Staff Duties
The following are the normal responsibilities of staff in a stable yard.

CHIEF INSTRUCTOR. Typical responsibilities:
Interviewing prospective clients and students.
Organising lessons and rides—checking daily ride-list in desk diary so that it is ready for the secretary to type.
Allocation of suitable horses.
Providing private lessons.
Preparation for exams.
Schooling of horses.
Co-ordinating with stable manager regarding lecture and practical stable-management sessions, and allocation of yard duties, etc.
Co-ordinating with secretary to keep records and book of rides and lessons.
Organising lectures and practical stable-management sessions for students and working pupils. These may be taken by a BHS Intermediate Instructor.
Arranging and conducting regular staff meetings.

STABLE MANAGER. Typical responsibilities:
Feeding.
Daily work lists.
Supervision of stable management of staff, students and working pupils.
Daily inspections of horses.
Arranging clipping, trimming, shoeing and veterinary list and saddlery repairs.
Attendance at veterinary surgeon's visits.
Feeding and checking of field horses.
Buying and storage of fodder.
Executing or arranging for necessary work in fields: i.e. harrowing, rolling, fertilising, topping, fencing, drainage, gates and water.

Executing and arranging general upkeep of stable yard, fences, gates, drainage, school surfaces (indoor and outdoor), show jumps and cross-country fences.

SECRETARY. Typical responsibilities:
Answering telephone.
Welcoming clients.
Booking rides and taking ride money.
Typing out daily ride sheets.
Supervising petty cash.
Dealing with correspondence.
Keeping accounts: VAT, wages and record books.
Liaising with chief instructor and stable manager with regard to record books and client problems or requests.
Keeping livery and ride accounts.
Sending out monthly bills.
Paying accounts.
Working out PAYE and employees' National Insurance contributions.
Putting up weekly wage packets or preparing wage cheques for signature.
Making regular visits to bank to pay in cash and cheques, and collecting wages and petty cash money.

INTERMEDIATE INSTRUCTOR. Typical responsibilities:
Looking after two horses.
Supervising stable management of students and working pupils.
Giving stable-management lectures.
Teaching.
Schooling.
General yard work.
NB: If preparing for BHSI, suitable help will be needed from the chief instructor.

ASSISTANT INSTRUCTOR. Typical responsibilities:
Looking after two to four horses.
Some teaching.
Joining in general yard work.
Helping to catch and groom field ponies.

NB: If preparing for an exam, lessons from the chief instructor and help from the stable manager will be needed.

WORKING PUPILS. Typical duties:
Looking after three horses.
Helping with beginner lessons.
General yard work.
Looking after and grooming field ponies.
Education should include:
Set periods of stable management each week: theory and practical.
Extra help as the examination date approaches.
Four riding lessons a week plus teaching practice.

PAYING STUDENTS. Work and instruction according to individual arrangements, but typical duties and education should include:
  Looking after one horse.
  Preparation for intended examination.
  Two daily lessons plus one lunge lesson.
  Teaching practice.
When sufficiently experienced, taking class lessons or beginner lessons under supervision, or lessons suitable in standard as preparation for intended examinations.

ODD-JOB MAN. Typical responsibilities:
Liaising with stable manager regarding daily work.
Stable repairs and yard maintenance.
Field work and fence repairs.
Grass cutting.
Building and maintaining show jumps and cross-country fences.
Care of car, horsebox and trailers.

In small yards, the duties of chief instructor and stable manager can be amalgamated, and some of their duties can be reallocated to an Intermediate Instructor or senior BHSAI. Smaller yards are less likely to have paying students, or to do exam preparation, and tend to concentrate on providing instruction and hacking for the general rider.

DOMESTIC STAFF. A cook, a housekeeper and cleaners may be required. If staff and/or students and working pupils are to live in or live in a staff house, arrangements have to be made for:

Cleaning the house.

Preparation of food.

Mature staff living in separate accommodation may be expected to look after themselves, but young students and working pupils should not be required to do so. It is the responsibility of the school owner or manager to see that young members of staff are well looked after. They should have suitable accommodation, and should be encouraged to care for it and keep it clean and tidy.

Turnout of Staff, Students and Working Pupils

Some yards will have strict rules regarding dress and turnout. All yards should make sure that personnel are suitably dressed. In particular, hair should be tidy; if long, it should be tied back or held in a scarf or net. The wearing of jewellery should be firmly discouraged. Ear-rings are dangerous.

When working in the yard, jeans, sweaters and gum boots are suitable. When riding, BSI standard head-gear and safe footwear are essential.

GENERAL DAILY ROUTINE

The following tasks are applicable to most stable yards:

1 Check the horse's general well-being:

That he is moving comfortably round the box.

That he is not pointing or resting a front foot — a sign of discomfort. If the opposite hind foot is also rested, it is a sign of relaxation after work.

That there are no signs of injury such as those caused by getting cast.

That the bed is in a satisfactory state and that the droppings are normal in appearance and number.

That the feed and hay have been eaten.

2 If necessary, adjust rugs.

3 Water, feed and leave horse to eat in peace.

4 Muck out, pick out feet into a skip, bed down, sweep up. If there is time, tidy up the muck heap.

5 Brush off and saddle up.
Horses should be tied up:
If the groom is working in the stable.
If they are saddled up ready for work, in case they try to roll.

6 Exercise.

7 On return from exercise, wash and pick out feet, unsaddle, give small feed of hay.

8 Groom.

9 Skip out box. Tidy muck heap (if not done earlier). Water, feed, lunch.

10 Finish grooming.

11 Clean tack.

12 Stable jobs (see *Stable Duties*, page 81).

13 Skip out.

14 Water and feed.

15 Sweep yard.

16 Give hay net.

17 Check:
Horse rugs.
Water.
Stable doors.
Feed shed.

18 Lock up tack rooms and close yard gate.

19 Later at night look round to ensure that all is well, but try not to disturb the horses.

20 Give water and late feeds if required.

Typical Daily Routine in a Riding School

Time	Work	Person Responsible	Lessons, Teaching, Schooling	Person Responsible
0700	Check horses. Water and feed. Horses working at 0800 have small feeds and are fed after work.	Stable Manager		
0800	Water. Muck out, bed down (or at 1215).	Staff	Private lessons	Chief Instructor
0900	Brush-off horses. Saddle-up for 1000 and 1100 lessons and for livery owners. Tidy muck heap and sweep yard.	Staff		
1000			Staff lesson	Chief I or II
1100			Client lesson	I or AI
1115	Small hay net all round. Groom horses.	Staff		
1215	Skip out. Water. Bed down.	Staff	Lungeing or schooling	Chief I or II
1230	Feed	Stable Manager		
1230	LUNCH BREAK			
1300			Private lesson	II
1345	Prepare horses for 1400 lesson.	Staff		
1400	Lectures for working pupils.	Stable Manager	Lesson	AI or II
1430	Catch and prepare ponies for 1600 lesson. Groom horses.	AIs and working pupils		
1500			Lesson	AI or II
1600-1730	Feed. Skip out. Hay up.	Stable Manager Staff	Childrens' lesson	AI and WP help
1730	Turn out ponies	WPs		
1800	HOME			
Evening Rides				
1700	Saddle-up for 1800 lesson. Prepare hay and leave outside boxes of horses being worked.	Staff on evening rota	Lessons	Staff on evening rota
1800-2100	Prepare horses. Skip out boxes. Water. Check unsaddling and rugging-up. Put away tack. Give hay nets.			
2100	Final check or after last lesson.	Stable Manager		

Typical Daily Routine for Competition Horses
This routine would be appropriate for a yard of three horses with one groom and help to ride.

0700 Check water. Feed.
0730 Muck out. Bed down. Sweep up. Tidy up muck heap.
0830 Brush off.
0900 Schooling and/or exercise.
NB: Hay after work.
1200 Strap one horse.
     Set fair others.
1245 Feed.
1300 LUNCH.
1400 Strap two horses.
1515 Stable jobs. Tack.
1600 Skip out. Water.
1630 Feed. Finish tack.
1730 Skip out. Water. Hay-up.
2030 Water. Fourth feed.

General Considerations in a Competition Yard
- Feed one and a half hours before exercise.
- *No* hay before exercise if working in the morning.
- If working early, give first feed after work.
- Check legs for heat or strain after work and in evening.
- If work reduced, cut down on the concentrate feed.

Typical Daily Routine for Hunter-liveries
This routine would be appropriate for a yard of four horses. When getting the horses fit, extra help will be required if other duties are to be efficiently performed.

0700 Check horses. Water. Feed.
0730 Muck out.
0830 Brush off. Saddle-up.
0900 Ride and lead each pair for approximately one and a quarter hours.
1015 Return from exercise. Prepare second pair and take out.
1130 Unsaddle. Hay net.

1145 Strap two horses. Skip out beds. Set fair. Sweep yard. Tidy muck heap.
1245 Feed.
1300 LUNCH.
1400 Strap two horses.
1500 Clean tack.
1545 Stable duties. Sweep yard.
1630 Feed.
1700 Skip out. Check rugs. Water.
1730 Finish.
2030 Check horses. Water.

General Considerations in a Hunter-livery Yard

• When the horse is fit and hunting twice a week, daily exercise can be reduced. In some cases three-quarters of an hour is sufficient.
• No hay should be given before hunting.
• A warm mash may be given after hunting, but this is no longer thought to be so beneficial as in the past. *See Book 7.*
• The day after hunting, check legs, trot up, lead out for ten minutes.
NB: One groom should be able to do five horses if he or she is not doing the exercising. A horse walker can be a great boon, and can save many hours spent exercising.

THE GENERAL STABLE DUTIES

Daily
1 Water.
2 Cleaning out water bowls.
3 Feeding.
4 Mucking out and sweeping yard.
5 Grooming.
6 Exercising.
7 Cleaning tack.
8 Washing feed bowls.
9 Tidying tack room and feed shed.

10 Washing sinks and floors.
11 Cleaning lavatories and basins.
12 Raking and harrowing school floor and track.
   NB: Raking done by hand; harrowing done by tractor.

Weekly
1 Dusting down stables.
2 Checking salt-licks.
3 Cleaning drains and traps.
4 Shaking and brushing rugs.
5 Disinfecting stable floors — not deep-litter bedding.
6 Washing lavatory floors and checking supplies.
7 Washing grooming kit.
8 Cleaning head-collars and rollers.
9 Oiling tack if it is wet.
10 Levelling school floor; salting and/or watering if necessary.
11 Dusting gallery.
12 Preparing farrier list.
13 Dealing with saddlery repairs.
14 Washing down barn and stable passageway.
15 Washing stable tools and wheel-barrows.

Monthly
1 Checking door bolts.
2 Checking string on tie-up rings.
3 Checking drains.
4 Ordering feed supplies.
5 Checking other supplies: e.g. disinfectant, saddle soap, clipping blades, brooms, stable tools.
6 Checking bandages.
7 Checking saddlery for repairs.
8 Washing stable walls.
9 Checking first-aid box and veterinary cupboard.
10 Cleaning out barley boiler.
11 Ensuring that horses are shod every four to six weeks.
12 Worming every six weeks.

Annually
1 Checking night and New Zealand rugs, washing and sending for repair.
2 Sending day rugs and blankets to be cleaned and repaired. Store away.
3 Creosoting wooden stabling and wood surfaces.
4 Painting block walls of stables.
5 Painting school walls.
6 Repairing stabling. Repaint as necessary.
7 Checking roofs and yard surface. Repair as necessary.
8 Checking windows.
9 Buying hay, straw and other bedding.
10 Checking saddlery for replacement.
11 Checking fields, fencing and gates before turning out horses for spring.
12 Checking fire extinguishers.
13 Servicing clippers, groomers, corn mill, water heaters and other equipment.
14 Organising inoculations.
15 Checking teeth every six to twelve months.

PREVENTION AND CONTROL OF FIRE

It is advisable to consult the local fire brigade headquarters about fire regulations relating to stables, and to seek their advice regarding a particular stable yard and its problem areas. For legal requirements, see Chapter 6, page 94.

Recommended Layout of Stable Yard and Surrounding Buildings

• Hay and straw barns should be well away from the stable area. Small amounts of hay and straw may be kept easily available for daily use in a convenient covered area.

• Tractor-sheds and horsebox garage should also be away from the stables and the hay barn.

• Petrol and diesel tanks are not in general use, but if you do have them on your premises they should be near the access road and away from all other buildings.

- If supplies of tractor fuel have to be stored, they should be in heavy-duty jerry cans. Not more than 10 gallons (45.5 litres) should be stored at a time. *Note*: Plastic containers are not safe for storing diesel fuel, petrol or paraffin.

Recommended Procedures and Practices
- No smoking in stable and working areas. 'No Smoking' notices should be prominently displayed and no-smoking rules strictly adhered to.

- Fire notices should be displayed in all stable blocks and in all other ancillary buildings. Instructions for procedure in the event of fire must be clearly set out and legible.

- Fire drill should be held at regular intervals, particularly when there has been a new intake of staff or working pupils.

- Suitable fire appliances appropriate to the contents should be placed in all buildings, including covered school and gallery.

- All personnel should be clearly instructed as to which appliances should be used on the following sources of fire:
  □ Hay and straw.
  □ Electrical faults.
  □ Oil and petrol.

- A sufficient number of fire alarms should be available in accordance with the size of the yard. They should be regularly tested to ensure that they are in working order. They should be protected from frost.

- There should be an adequate number of hoses and water-points, so that the hoses reach all areas of stables, barns, and other buildings.

- Unless there is a risk of frost, the main fire hose should always be kept fitted to the water supply, and ready for use. In freezing weather, it should be drained and insulated.

Action to be Taken in the Event of Fire (These instructions should be given to all personnel).
1 Sound the fire alarm.
2 If the stables are on fire, or could catch fire, take or release the horses into a field or designated area. Begin with those nearest to the fire.
3 Call the fire brigade, or make sure that someone else has called them.
4 Tackle the fire with appropriate fire appliances.
5 Make a head count of all staff and other persons on the yard.

Horses may be frightened and refuse to leave their stables. A coat or cloth put over the head and covering the eyes may make them more willing to move. In the case of a difficult horse, several people may be needed to move him. Inhaled smoke is very dangerous to horses, so make sure that they have access to fresh air. If they *have* inhaled smoke, it is advisable to seek veterinary advice.

The Most Common Causes of Fire
- Electrical fault in wiring or appliance. The installation of a trip-switch, which cuts off the electricity supply in the event of a fault, should avert this.

- Electric or paraffin heaters.

- Clothing or rugs put too close to heaters when drying.

- Smoking. Careless use of matches and throwing away of cigarette ends.

- Children playing with matches.

- Self-combustion of hay in the stack.

- Bonfires placed too close to hay barns or other buildings.

CHAPTER 5
Buying Fodder and Bedding

POSSIBLE ARRANGEMENTS

☐ A verbal contract can be arranged with a reputable corn merchant for regular deliveries throughout the year. This system is recommended for yards with limited storage space, or for inexperienced buyers. The latter have to rely on the good faith and reputation of the merchant to ensure that they receive goods of high quality. It should be agreed in advance that if any delivery falls below standard, the goods will be replaced at no extra expense to the buyer. This arrangement is likely to be more expensive than buying from a local farmer, but should ensure guaranteed good-quality supplies throughout the year.

☐ A verbal contract can be made with a local farmer to supply a given quality of fodder and straw throughout the year. Deliveries should be by arrangement, and as convenient. Prices are cheaper than when buying from a merchant. The quality of hay, straw and corn should be checked at harvest time, and again before delivery, to make sure that there has been no deterioration.

☐ Bulk buying can be arranged direct from the field, thus avoiding extra handling, transport and storage costs.
   The practice of buying from a neighbouring farmer has much to commend it. There will be many occasions

during the year when goodwill on both sides is needed. It is important to be on friendly terms.

☐ Manufactured feedstuffs have to be bought either through a corn merchant or direct from the manufacturer. When bought by the tonne or tonnes, a considerable discount should be obtained. Bought by the sack, goods can be collected and paid for on a cash-and-carry basis.

Bulk Buying

It is always cheaper to buy in bulk, rather than in small quantities. Suitable storage facilities with a sound access surface are essential, as is the cash necessary to pay for the goods. If money is short, it is often possible to borrow from the bank; even with interest added, this may still make a considerable saving.

BUYING OF HAY

☐ If hay is bought off the field at harvest time, extra handling, transport and storage costs are avoided. Even if it is bought at a later date, but in sufficiently large quantities, a favourable price can usually be negotiated.

☐ If bought in sufficient quantity, a continuing supply of good-quality hay is assured. There is no sudden change in quality content and, with good management, enough old hay will be left to carry on until the new season's hay is ready for use.

☐ According to the condition of the hay when baled, a minimum weight loss of 25 per cent may have to be allowed for. This occurs as the hay dries out.

Buying of Straw

The same points apply as with the buying of hay except that there is little weight loss as straw is dry when baled.

STORAGE OF HAY

Suitable storage is essential. Poor or inadequate facilities mean hay wasted and money lost.

Purpose-built Hay Barn

This is the most expensive but also the most satisfactory method. The necessary money can often be borrowed from the bank, and the yearly saving in hay costs should make it worthwhile. The barn should have a ridged roof, with air vents to allow ventilation at the top of the stack and the escape of fumes from the new hay. If sufficient money is available, the north and west – or most exposed – sides of the barn should be filled in with slatted boards as a protection against snow and rain. The two most convenient sides should be left open to facilitate the unloading and stacking of the bales.

Conventional Barn of Brick or Stone Construction

This usually holds only a small quantity and is often more awkward to stack because of enclosing walls; however, the sides cannot fall out even if stacked by inexperienced workers. Although this type of building is useful if already in the yard, it may lack ventilation, and there will thus be more chance of the hay heating up if it has not been sufficiently 'made'.

Covering for an Open Stack

Plastic sheeting.
Canvas rick sheet.
Both of these can be successfully used to cover the stack, but there is more chance of hay and straw being wasted.

Building the Stack

THE BASE. To avoid spoiling the bottom bales of a stack, it is usual to put down a base of old hay or straw. This can be loose or in bales, but must be thick enough to prevent any damp getting up into the new hay. Small amounts of hay (10 tonnes or less) can be stored on wooden pallets. These are not practical for the larger stacks because of the weight of hay.

BUILDING THE STACK IN A PURPOSE-BUILT HAY BARN. If there are no outside walls, the stack must be built with great care. One bay should be filled at a time. The front should be filled first, and then the sides and back. The middle should be filled last. The bales should be positioned so that each layer locks the layer below and holds it steady. This work requires experienced workers or careful supervision. Should there be any tendency for the side of the stack to bulge out, this can usually be checked by supporting the sides with long poles (jump poles will do). The same principles apply when loading hay or straw on to a lorry or trailer. If transporting it on a road, always rope the load for greater safety.

Heating in the Stack
If carted from the field immediately after it has been baled, hay heats in the stack. To avoid overheating, it is advisable to allow for air passages in each bay. Stand a bale upright at base level in the centre of the bay. As the layers of hay are stacked, pull up the bale with them, thus leaving a square ventilating shaft. At the top of the stack, leave a space between the top layer and the roof so that air can move freely. If the weather looks settled, hay often does better if stood up in the field for a few days before carting.

Stack to be Covered by Plastic Sheet or Canvas Rick Sheet
Build the stack as described above, but with even greater care, as there are no uprights or boards to support the sides. The hay is stacked in a square or rectangle with a flat top. The sheet is put over the top, and should be large enough to hang over the edge. If heating is likely to be a problem, build accordingly (see *Heating in the Stack*, above), but also put six bales on top of the last layer. When the sheet is placed over the bales, air can circulate underneath. When heating has finished (two or three weeks), the stack can be secured for the winter. Remove the six bales so that

the sheet lies nearly flat; a very slight slope is a help to chute the rain. If using a plastic sheet, place spare bales or heavy-duty tyres on the top to hold it down. This is not necessary with a canvas sheet. The sides are tied down with rope or nylon string, or pegged to the lower bales. This arrangement should be weatherproof. Once it becomes necessary to start using the hay or straw, there is always the problem in resecuring the sheet – and wastage and spoiling of either hay or straw ensues.

Plastic sheets are cheap, and with a stack in a sheltered area and treated with care, they can be a success. If they are not well secured, or if the weather is very rough, they rip and tear very easily: so they are not suitable for windy districts. They usually last only for one year. Large-mesh nets can be bought or made out of baler twine. If these are put over the sheets they give added security and reduce the chance of tearing.

Canvas rick sheets are more expensive, last longer, and are easier to secure to the stack. When the stack is opened, they are easier to replace, and stand much more rough handling and rough weather.

Opening the Stack
This should be carried out by experienced staff, but if they are not available, the workers must be well briefed and supervised.

When opening a stack, take hay from the top and work down. If individual bales are pulled out from the bottom this can unbalance the rest of the stack, which may fall and injure anyone standing below. It will also spoil many of the bales. When removed from the stack the bales should be stored in a hay shed adjacent to the stables where they will be ready for use.

Storage Areas and Quantities
Hay and straw can be bought either by the tonne or by the bale. When buying by the tonne:

Each load can be weighed on a weighbridge. This is costly in time and fuel.

or

10 to 20 bales can be weighed, and from that an average weight per bale can be worked out.

A barn of dimensions 45 ft × 30 ft × 18 ft (13 m × 9 m × 5 m) houses approximately 100 tonnes of hay or 4000 bales.

100 tonnes of hay is sufficient for approximately 50 stabled horses for a year.

65 tonnes of straw is sufficient for approximately 50 stabled horses for a year.

Hay bales can weigh 30 to 60 lb (13 to 26 kg). Straw bales can weigh 30 to 60 lbs (13 to 26 kg).
The weight of the bale depends on:
The condition of the hay when baled.
The setting of the baler. This can be altered to make light or heavy bales. An all-female staff will appreciate the lighter bales, even if there are more of them to the tonne.

BUYING BALED SHAVINGS AND PAPER BEDDING

These are both much cheaper when bought by the lorry load. They can be stored outside without a cover, as they are packed in polythene bags. Choose a level site convenient to the stables.

LOOSE SAWDUST AND SHAVINGS

If you are in a wooded district, these (according to demand) can be:
1 Free. A verbal contract is agreed whereby all available supplies will be regularly collected.
2 Delivered by lorry. Transport has to be paid for.
3 Bought by the sack. Provide your own transport or pay for delivery.

BUYING AND STORAGE OF GRAIN

Bulk buying of grain presents more problems, and is not really practical for the small yard. The problems of good quality and an assured and regular supply can often be solved by contracting to buy an agreed amount from a local farmer. He will be responsible for storage and delivery as requested. Payment can be arranged monthly. Corn is always in short supply by July, and there can be a problem until the new grain is ready to feed. It is important to remember this when buying in bulk earlier in the year, or when making arrangements for regular deliveries. Make sure that the supplier knows the approximate total quantity needed for the year.

If keeping high-performance horses, this arrangement may not be satisfactory, as the necessary quality of grain may not be available locally. Under such circumstances, and if grain is being bought to last for the year, a silo capable of storing sufficient grain for twelve months can be the most practical proposition.

Before storing grain in bulk, moisture content should be checked. Damp grain – i.e. with moisture content of over 14 per cent – will heat and spoil. In a good harvest year, or in the case of imported grain, there is no need to dry the grain before storing. In a wet season, grain can have as much as 21 per cent moisture, which must be reduced by drying before storing. Drying affects vitamin and mineral content, and this has to be taken into account when working out daily rations.

Quantities of up to 4 tonnes can be stored in purpose-built metal containers. Alternatively, a spare brick or concrete-block loose box with a solid door can be used as a store. The roof must be made vermin proof.

Whole corn can be bought in small quantities and stored in metal bins, or paper or jute sacks (not plastic). Living seed dries naturally, and will not deteriorate or diminish in food value. If losses are to be avoided, the storage area must be vermin proof or well patrolled by cats.

Bruised, rolled, crushed or cut grain should not be bought in bulk. It is dead, the husk is broken, and its food value deteriorates within a week. Grain is often dampened during processing, which can cause it to heat and spoil if not fed within days.

CHAPTER 6
The Law

This chapter outlines the relevant contents of the Acts which cover horses and stables. The Acts discussed are the *Protection of Animals Act of 1911 and 1912*, the *Animals Act of 1971*, the *Riding Establishments Acts of 1964 and 1970* and the *Health and Safety at Work Act of 1974*. Only outlines can be given, and they are not a substitute for professional advice.

RIDING ESTABLISHMENTS ACTS 1964 AND 1970

The setting up, and to a large extent the running, of a riding establishment, is governed by the combined effect of the *Riding Establishments Acts 1964 and 1970*. These are applicable in England, Scotland and Wales.

Northern Ireland has separate legislation entitled *Riding Establishments Regulations 1980*, and riding stables in the *Isle of Man* are governed by the *Riding Establishment (Inspection) Act of 1968*.

The Riding Establishment Acts 1964 and 1970 forbid the keeping of a riding establishment in England, Scotland and Wales except under the authority of a licence issued by the local authority in whose area the premises (including land) are situated.

THE RIDING ESTABLISHMENT

The term Riding Establishment covers the carrying on of

a business 'of keeping horses to let them out on hire for riding and/or for use in providing instruction in riding, for payment'.

Licences
A licence may be granted annually after application by an individual over the age of eighteen years, or by a company. The normal duration of a licence is twelve months. Licences are not renewable, and a new application and inspection of the premises is required on the expiry of the existing licence. Licences run from the date of issue, but some local authorities issue licences dated from 1 January.

As an alternative, if a local authority is not satisfied that a case has been made for a full licence it has the power to grant a provisional licence for a period of three months. This provisional licence can specify the conditions required, so that the licence holder has time to meet such conditions and will therefore eventually be granted a full licence. The local authority may extend the period for a further three months, but only on re-application by the licence holder. One of the purposes of such an extension is to give the licence holder the opportunity to complete work already begun in order to meet the specified conditions. The local authority may not issue provisional licences to any person or company for more than six months in any one case.

The cost of the licence, either full or provisional, is at the discretion of the local authority, and the total fee includes administrative and inspection charges. No provision is made for a charge to be levied for the three-months' extension to a provisional licence.

The Acts give the local authority complete discretion over the granting or refunding of a licence. However, an aggrieved applicant can appeal to a magistrates' court, in relation to both the refusal and to any conditions that the authority has imposed. The magistrates may then give such directions as they think proper in respect of the issuing of a licence or to the conditions.

There are no provisions within the Acts for third parties to appeal against the granting of a particular licence.

Planning Permission

Before consideration is given to the granting of a licence to a new applicant, the local authority should ascertain that under the *Town and Country Planning Act of 1971* permission has been obtained either to erect new buildings, or for change of use of all or part of existing premises.

Qualification for a Licence

No-one under eighteen is qualified to apply for a licence, nor is any person who for the time being is disqualified under any of the following:

- The Riding Establishment Acts.

- The Protection of Animals (Cruelty to Dogs) Act (1933) from keeping a dog.

- The Protection of Animals (Cruelty to Dogs) (Scotland) Act (1934) from keeping a dog.

- The Pet Animals Act (1951) from keeping a pet shop.

- The Protection of Animals (Amendment) Act of 1954 from having custody of animals.

- The Animal Boarding Establishments Act (1963) from keeping a Boarding Establishment for animals.

An applicant for a licence must be a qualified person within the above provisions, and he/she must be able to satisfy the local authority that he/she is suitable, and qualified, either by practical experience in the management of horses or by being the holder of one of the following certificates: BHS Assistant Instructors, BHS Intermediate Instructors, BHS Instructors, BHS Fellowship, A.B.R.S. Fellowship, or any other certificate prescribed by order of the Secretary of State. An inspecting veterinary surgeon is required to consider the suitability of the applicant in accordance with the above conditions.

Inspections

Before granting a licence the local authority grants powers in writing to an officer from their own or from any other local authority, a veterinary surgeon or veterinary practitioner selected from the list specifically drawn up jointly by the Royal College of Veterinary Surgeons and the British Veterinary Association for this purpose. He/they should carry out a detailed inspection of the premises and submit a report to the local authority, who will consider whether the premises and the persons employed in the management of the riding establishment are suitable to be holders of a licence.

Inspectors are recommended to visit the riding establishment at any time within reason when all horses are likely to be at the stables: giving not more than 24 hours' notice of the visit. The inspection should be made at the time of year appropriate to the type of use of the animal, and to seasonal activities such as trekking or hunting. In order to determine the standard of care and management, the animals should be in full use.

Conditions of a Licence
Without prejudice to their discretion to withhold a licence on any grounds, the authority shall take particular regard to the following:
• Accommodation should be suitable for the animals in respect of size, construction, number of occupants, light, ventilation, drainage and cleanliness. Stalls should be wide enough and long enough to allow the animals to lie down and to get up easily and without risk of injury. Boxes should be large enough to allow the animal to turn round. Stalls and boxes should be free from fittings, projections or structural features which might cause injury. Doors should always open outwards. These and the following requirements apply to new constructions, and to buildings which have been converted for use as stabling.

The local authority considers the number of horses kept at an establishment (including animals at livery) in relation to the buildings and land available, and may impose a condition specifying the maximum number of horses of all categories (both for use in the riding establishment or otherwise) which should be kept at any time. Yards should provide sufficient space for every animal kept on the premises.

- Lighting should be adequate enough to preclude the use of artificial light during daytime. Switches, wires and other electrical equipment should be protected in such a way that horses cannot injure themselves.

- There should be adequate ventilation without draught.

- Drainage must carry away any liquids voided, and be sufficient to keep the boxes and stalls dry. Provisions must be made for the storage and regular disposal of manure and waste bedding.

- Adequate and suitable food, water, and bedding, together with both rest and exercise where required, must be provided, as well as suitable facilities for the storage of reasonable reserves of food and bedding.

Grazing
Where horses are kept at grass there must be adequate pasture, suitable shelter, and water at all times. When the animals are either in work, or during the winter period when the grass is not growing, the Acts require that adequate supplementary feed should be given. The veterinary surgeon takes into consideration the type of animal, together with the type and location of their pasture. Arabs, Thoroughbreds or Hunter types require more protection from the weather than do native ponies.

It is necessary within the Acts to maintain fences in a safe condition and to keep the grazing free from hazards and rubbish. A competent person who can recognise injuries or illness must visit the horses at grass daily.

Horses
☐ The term 'horse' within the Acts includes any mare,

gelding, pony, foal, colt, filly, stallion, ass, mule or jennet. They are required to be kept in good health, physically fit, and suitable for the purpose for which they are being maintained.

☐ It is not permissible to use animals three years old or under, or mares heavy in foal or within three months of having foaled. Good, not thin condition is required. It is not necessary for horses to be shod, but their feet must be kept well trimmed and in good condition. If they *are* shod, the fitting of the shoes must be correct.

☐ Animals must be free from illness, sores, galls, or injuries from the bit, saddle or other sources. Where injury or illness has occurred it is an offence not to provide curative care and treatments.

☐ On any inspection of the premises by an authorised officer of the local authority, a horse found to be in need of veterinary attention must be removed immediately from work on the verbal instruction of the officer. These instructions are generally confirmed in writing as soon as possible after the inspection. The horse cannot be returned to work until the licence holder has obtained at his/her own expense a certificate from a veterinary surgeon that the horse is fit for work, and has lodged the certificate with the local authority. This is a mandatory condition in the granting of a licence.

☐ A register of all horses three years old and under and in the possession of the licence holder must be maintained and must be available for inspection.

☐ Horses at part livery and partially used in the riding establishment are within the provisions of the Act and subject to inspection.

☐ Horses kept at full livery for private owners should be noted and can be inspected by the authorised officer who has powers under Section 1 (3) of the principal Act to inspect any horse found on the premises.

The Law

☐ It is an offence under the Acts to conceal, or cause to be concealed, any horse maintained on the premises, with the intention of avoiding an inspection of this horse.

☐ It is also an offence to allow a horse to be in such a condition that riding him would be likely to cause him suffering, as would letting him out for hire, or using him to provide, in return for payment, instruction in riding, or to demonstrate riding, whether for payment or not.

☐ When instruction in riding is given on a horse which is the property of the pupil receiving instruction, it is not usually considered that this requires a licence, regardless of where that instruction is given.

Saddlery

All riding equipment must be maintained in good condition and, at the time when it is supplied to the rider, should not be subject to any defect which on inspection is considered to be likely to cause suffering to the horse or an accident to the rider. It also constitutes an offence to supply a saddle which is ill-fitting, causing a sore back; or bits, curbs, etc. which cause injury to the mouth; or bridles, girths, stirrup leathers and irons which might break due to faulty materials, manufacture, or stitching and thus place the rider in peril or the horse liable to injury.

The inspector examines the saddlery when fitted to the horse for which it is intended, and pays particular attention to the correct fitting of Western-type saddles and also to ex-army troop saddles, as many of these are not fitted with safety stirrup bars and might cause a falling rider to be dragged. This type of saddle should be used only with patent safety stirrup irons.

Infectious Diseases and First-aid Equipment

● All reasonable precautions must be taken against the spread of infectious diseases and, so far as it is possible, there should be provisions to isolate an infectious animal.

• Veterinary first-aid equipment and medicines must be kept on the premises in a suitable and clean place set aside for this purpose. It is recommended that before assembling this equipment, consultation as to its contents should be made by the licence holder with the establishment's veterinary surgeon. He will take into consideration the veterinary experience of the licence holder when advising on particular medicines, etc., as well as on the standard contents, such as antiseptic solutions, powders, bandages, dressings, scissors and a clinical thermometer. It is also strongly recommended that the name, address and telephone numbers of the establishment's veterinarian and doctor should be prominently displayed.

Fire Precautions

Precautions must be taken for the protection of the horses in case of fire. Notices prohibiting smoking should be displayed.

There must be clear access to all stalls and boxes, and where multiple numbers of stalls and boxes are housed within the same building, more than one exit is strongly recommended. Fire extinguishers must be regularly serviced, and an adequate supply of water should be available easily. The local fire-prevention officer should be consulted on the types of fire-fighting equipment most suitable for the premises.

The *1970 Riding Establishments Act* requires that there should be a notice on the outside of the premises, giving the name, address, and telephone number of the licence holder or other appointed responsible persons, together with directions as to the actions to be taken in case of fire. The directions should include particular advice on the removal of animals from the stables.

Management and Supervision

A person under sixteen years of age must not be left in charge of the business. All horses let out on hire must be supervised by a person of sixteen years or over, unless the licence holder is satisfied that the hirer is competent to ride without supervision. Before he can be satisfied that the rider is competent to ride unsupervised, a licence holder must therefore have either previous knowledge, or have made an assessment of the capability, of the rider, and must then provide a horse 'suitable for the purpose' of that rider's requirements and ability. A licence holder who has failed to make the correct assessment might be unable to use this as a defence should litigation follow.

The knowledge of the licence holder is an integral part of the granting of a licence, and should this knowledge be regarded as inadequate, the applicant can appoint a manager or other person to supervise the riding establishment on his/her behalf. The local authority can then issue a licence which will also carry the name of the person with the necessary knowledge or qualifications upon it. If this person leaves the establishment during the currency of the licence, it would appear that the licence becomes null and void.

Similarly, it would appear that a riding establishment cannot be sold as a licenced business, because without the qualifications or experience of the licence holder the licence is ineffective. Any purchaser therefore has to make an application to the local authority for a licence before proceeding to operate the business, and to satisfy the authority that he/she has the necessary qualifications or experience.

Provisions are made whereby in the event of the death of a person holding a current licence, the licence passes to his personal representatives, and can be extended for one year (three months if a provisional licence) from the date of death. Such an extension can be made more than once if the local authority is satisfied that the extensions are necessary to wind up the estate.

Insurance
The *1970 Riding Establishments Act* clearly states that the licence holder shall hold a current Public Liability Insurance Policy which provides indemnity against liability at law to pay damages for accidental bodily injury, sustained by the hirer of a horse, or those using the horse to receive

instruction in riding. The licence holder must also insure the riders in respect of any liability which they might incur through injuries to any other person, caused or having arisen through the hire or use of the horse.

The amount of indemnity is not specified, but licence holders are advised to make sure that the amount exceeds the highest awards the courts have made in respect of riding accidents.

In addition, where the business employs staff, compulsory insurance is required under the following:
☐ The *Employer's Liability (Compulsory Insurance) Act 1969.*
☐ *Employer's Liability (Defective Equipment) Act 1969.*

Rights of Local (Licensing) Authorities
● Powers of entry.
● The right to impose conditions upon a licence in order to secure all the objects specified within the Acts.
● The right to dispense with some conditions if they think that circumstances so warrant.

Five conditions in Section 2 of the 1970 Act are mandatory. They are:
1 The removal of horses from work which are in need of veterinary attention.
2 The supervision by responsible people aged sixteen years or over.
3 Leaving the premises in charge of a person over sixteen years.
4 The insurance policy requirements.
5 The register of horses three years old and under.

Any person who operates or has intentions of operating a Riding Establishment is advised to obtain copies of both the 1964 and 1970 *Riding Establishment Acts* in order to be fully conversant with their requirements.

THE HEALTH AND SAFETY AT WORK ACT 1974

The provisions of this Act apply to employers and

employees who are engaged in the keeping and management of livestock. Riding schools, livery yards, etc., are not specified, but do come under the Act. Responsible employers should be aware of their duty, both to themselves and to their employees, to see that work conditions are as safe and as healthy as possible.

Any business employing more than five people, including casual workers, must issue a Safety Policy Statement. Instructions to employees should be clearly set out. Assistance in drawing up this statement can be obtained from the BHS Riding Schools Approvals Office, BEC, Stoneleigh, Warwickshire; or The Association of British Riding Schools, Old Brewery Yard, Penzance, Cornwall TR18 2SL; or the local Health and Safety Executive (address and telephone number in the local Yellow Pages Directory). The local office of ADAS also supplies leaflets and a poster on Safety Measures and safer working conditions.

In December 1986, HM Agricultural Inspectorate of Health & Safety published requirements extra to the 1974 Act. Many of the following directives are covered, but in addition there are strict regulations for health and safety in relation to food preparation, first aid, sanitation, etc. Responsibility for staff training, instruction, suitability of horses, dress, etc, are also covered. It is therefore essential for persons setting up – or running – a yard to obtain a copy of *Horse Riding Establishments Guidance on Promoting Safe Working Conditions* by the Health and Safety Executive. They can thus ensure that they are able to comply with the necessary requirements of HM Agricultural Inspectorate.

General Obligations
- Employers must ensure the safety of their employees by maintaining safe systems of work, safe premises, and safe equipment.
- Employees and the self-employed must take reasonable care to avoid injury.
- Employers, the self-employed and employees must not

endanger the health and safety of third parties.

☐ There should be a named person to whom any faults in equipment or other hazards can be reported.

☐ Employers must ensure that all employees and others on the premises are correctly instructed in any work that they have to do and any equipment that they have to use.

☐ Employees should be encouraged to produce new ideas for improved safety measures and methods of working.

☐ Employers should ensure that their insurance policies cover the use of such equipment and machinery by any member of staff, paid or unpaid.

☐ Employer's liability insurance is compulsory for all employers, and an up-to-date certificate of this must be displayed.

☐ Well-equipped first-aid facilities should be available on the premises and a responsible person always on call when required. Leaflets on first aid can be obtained from HMSO Publications Centre, P.O. Box 276, London SW8 5DT or the local Health and Safety Executive.

Risk Areas
- Employers should give these special attention.
- Employees should be instructed as to any special precautions which should be taken.
- Employees should be instructed in the correct and safe use of equipment and machinery.

Employees likely to be left in charge of the yard, in the absence of senior staff, should be fully briefed as to what to do should an emergency arise. A list of relevant telephone numbers – for doctor, veterinary surgeon, owner or manager, electrician – should be provided.

Employees should be instructed about:
☐ The working of the trip-switch and how it is re-set.
☐ The whereabouts of all water-main stop-cocks.

105

The Law

The main risk factors are:

- COMBUSTIBLE MATERIALS. Stables. Hay barns. Oil, diesel or petrol storage areas. Electric, gas or oil heaters. Bonfires.

- ELECTRICAL EQUIPMENT AND MACHINERY such as clipping and grooming machines, corn mill, etc. All electrical wiring, fittings and equipment should be regularly inspected. Equipment should be regularly serviced. A circuit-breaker plug should be used with all electrical equipment. Electric cable should be inspected before using any equipment, to see that it has not been damaged by friction or by a horse treading on it. Plugs should be checked for faulty wiring and cracked casings. Personnel should be properly instructed in the equipment's use and care. Junior staff should be supervised when using electric clippers and groomers. The corn mill should be worked only by experienced staff. If it is not fitted with a dust extractor, face masks must be used. The dangers of inhaling dust should be explained.

- CHAFF CUTTER. If electrically powered, see previous paragraph. If either hand or electric powered, guards must be fitted to protect operator's hands. Warning notices should be fixed on the machine, and the risk of getting hands or clothing caught up should be explained.

- BARLEY BOILER. If electrically powered, see *Electrical Equipment*, above. If gas powered, it should be well away from any combustible material, on a concrete floor and be worked only by experienced personnel.

- TRACTORS, FIELD MACHINERY, HAY TRAILERS, ETC. Regulations and advice are published in a Ministry of Agriculture, Fisheries and Food pamphlet, which should be essential reading for all personnel.
  No extra persons should be allowed to travel on the tractor, hay trailer, or other farm vehicles.
  Personnel driving the tractor should have a current driving licence.

Tractors must be suitably insured, and if driven on or across a road, the necessary road tax must be paid. The dangers of working on steep hills, close to deep ditches and boggy areas should be explained; also the risk of allowing children to play in the area where the tractor is kept or where it is working.

- LAND-ROVER, LAND-ROVER AND TRAILER, HORSE BOX. Drivers must check that children and animals are not near when drawing away or reversing.

- HORSE WALKERS OR HORSE EXERCISERS. These must be well maintained in a suitably fenced area. They should only be used by experienced personnel. Horses should not be attached to the machine by junior staff.

- OUTSIDE STRUCTURES. Buildings should be kept in sound order. In strong winds, loose tiles, slates or guttering can be dislodged and cause injury to anyone passing underneath.

- STEPS AND STAIRS up to or down into working areas should be sound and with a level tread. It may be advisable to fit a hand-rail or rope.

- LADDERS AND STEP-LADDERS must be strong enough for the work involved. Unsupervised junior personnel must not be allowed to use ladders. Two people should be responsible for putting up the ladder and for taking it down and putting it away. Ladders should never be left up against a stack or near a stack, as they can be knocked down, breaking the ladder and possibly causing injury. They may also be an invitation to children to climb up and play on top of the stack. This must never be allowed.

- LIFTING LOADS
  - ☐ When collecting hay from a stack, employees must be instructed to take bales from the top of the stack and work down. Lower bales should not be pulled out, as this may cause the stack to fall, resulting in serious injury.

☐ When loading hay bales on to a trailer and taking them to the yard they must be stacked with care, so that the load is steady. If it is to be transported on a public road, the load should be roped for extra safety. Extra care should be taken on a steep or uneven surface.

☐ When lifting bales and sacks of feed, personnel must be instructed to lift weights in the proper manner. If in doubt, they should ask for help. Senior staff should check that the weights are not too heavy for the persons involved. They should not be allowed to risk straining their backs.
Instructions for lifting sacks and other heavy weights:
Estimate the weight and if necessary ask for help.
Stand close to the sack.
Square the sack up in front of you.
Bend your knees; do not lean over and lift up.
Take hold of the sack. Lift it by straightening the legs.
Look where you are going. When putting the sack down, bend your knees. Try not to bend your back.

● YARD AREA. Stable tools must be carefully used and tidily replaced after use. If left about they are a hazard to staff and horses: they can fall down, be tripped over or stepped on, resulting in a cut foot and/or bruises to face and eyes. Rakes should always be left standing upright, with the head at the top.

● BALING STRING. When the bale is to be used, the string should be cut, tied up and placed in a suitable sack or bin. Left on the ground, uncut, it can cause a fall and injury should either a person or a horse get their feet caught in it.

● WHEEL-BARROWS should never be left about. When in use, they should be placed parallel to the stable wall. Handles should never be left pointing outwards to the yard. When not in use, barrows should be parked tidily in the stable tool area.

- YARD SURFACES should be swept clean. It may be necessary to *hose* them clean should a film of mud and manure form, as this can be very slippery. It is likely to happen in the spring of the year when horses and ponies are brought in from the fields.

- NARROW DOORS AND GATEWAYS should have warning notices. Internal swing doors for feed shed, tack rooms or office are dangerous and are not recommended.

- COLD WEATHER CONDITIONS. In freezing weather salt should be used if there is ice on the yard. (This should be obtained in the autumn.) In deep snow, working paths should be cleared and then kept clear with regular applications of salt. As an emergency measure, sand or coal ashes help to keep a slippery surface usable. In continuing freezing weather, straw or shavings manure tracks, six feet wide, give a secure footing for horse and staff. This is more effective than using plain straw.

  Water buckets must be emptied carefully into a drain, and not spilt on the yard to create an icy surface.

- FIELDS, FENCES AND GATES should be kept in good order. Heavy gates on broken hinges can cause a fractured leg for human or horse. Bog and deep ditch areas should be fenced off. If a field entrance on to a road is dangerously placed it should preferably be moved but, if this is not possible warning notices should be fixed, and staff should be advised of the danger.

- CAR PARK, AND YARD AND DRIVE ENTRANCES. The following warning notices should be fixed:

  ☐ Speed restriction 5 mph.

  ☐ Children and horses.

  Drives and access roads can be made safer by placing 'humps' at intervals to slow down traffic.

Cars, horseboxes and trailers should be parked away from the stable yard, preferably in a car park. There should also be a special area set aside for bicycles.

- TREES. Tall old trees near the stable yard and car park area can be a hazard. They should be regularly inspected, and if necessary lopped or felled.

THE ANIMALS ACT (1971)

This Act is not applicable in Scotland or Northern Ireland. It is based on strict liability and deals with the injury that horses may cause, and with damage arising from their straying.

Strict liability means that there is liability without proof that the person claimed against was negligent, if he was responsible for the animal, and the animal has caused injury or damage.

Animals

These are divided into animals belonging to a dangerous species, and those which do not. Bulls and wild and unbroken stallions may come within the former category. Liability for injury caused by animals of a dangerous species will be placed on the 'keeper' unless he can bring himself within one or more of the exceptions of the Act. Liability for damages caused by an animal which is not of a dangerous species (such as a horse or pony) also rests with the 'keepers' in the following circumstances:

☐ The damage is of a kind likely to be caused, and when caused is likely to be severe, unless the animal is restrained; and the likelihood of the damage, or of its being severe, must be due to the characteristics of that particular animal which are not normal in animals of this species except at particular times or circumstances: e.g. mares in season, etc; and those characteristics were known to the keeper or known to any other person who had charge of the animal as the keeper's servant, or where that keeper is the head of a household, or were known to another keeper of the animals, who is a member of that household and under sixteen years of age.

Who is a Keeper?
A person will be treated as a keeper if:

☐ He owns the animal, or has it in his possession, or;
☐ He is the head of the household of which a member under sixteen years of age owns the animal or has it in his possession.

The keeper must be an individual. The person who has control over company-owned horses would be regarded as the keeper. Liability for damage rests on the person who at the time has possession; he need not be the owner. Anyone who takes possession of and keeps horses to prevent them from causing damage or to restore them to their owner is NOT a keeper.

Damage by Trespassing Stock
In broad terms an animal is said to trespass in the same way as humans: that is, where it is on land on which it has no right to be, or where its owner has no right to put it. This includes animals straying on to a highway or from a highway on to private land. Exceptions exist with regard to animals straying from unfenced common land on to highways.

The only lawful use of the highway (unless they have grazing rights on the verges) is to pass and repass on it. The highway need not adjoin the land; the stock may have wandered through other land from the highway. Bridlepaths are highways.

Section 8 of this Act imposes a duty on the person placing horses on the land to take care to prevent them from straying on to the highway: the only exception being if they stray from unfenced land, and providing the person who placed them on the land had a 'right' to do so. Unfenced land is regarded as land in an area where fencing is not customary; this is common land, town or village green or some moorlands.

The Law

Claims under Section 2 for Injuries and Damage Done by Animals

1 Where a horse strays on to land owned or occupied by another person, the owner of the horse is strictly liable for the following damage and expenses:

(a) Damage done to the land or to any property on the land which is owned or in possession of another person.

(b) Expenses which are reasonably incurred by the other person in keeping the horse until it can be returned to the owner.

(c) The expense of finding out to whom the horse belongs.

(d) Expenses incurred by the occupier of the land in exercising his right to detain the horse.

(e) Injuries caused by the straying horse; death of or injury to, a person is included in the word 'damage'.

2 There would be NO responsibility for straying live-stock if:

(a) The damage is due wholly to the fault of the person suffering it, but that person cannot be regarded as at fault just because he could have prevented it by fencing, or

(b) It is proved that the straying would not have occurred but for a breach of duty imposed on another person who has interests in the land to fence it, or

(c) The person suffering the damage has voluntarily accepted the risk. But where a person employed by a keeper incurs a risk as a result of his employment he shall not be treated as accepting it voluntarily.

(d) The livestock strayed from the highway whilst lawfully using it.

3 The occupier of the land on to which the horse strays without being under the control of any person has the right to detain it, unless ordered by a court to return it. The right ceases, however:

(a) At the end of 48 hours unless the occupier notifies a Police Officer and also the person to whom the horse belongs, if that person is known.

112

(b) If the person claiming the horse offers the occupier sufficient money to cover any proper claim for damage and expenses.

(c) If the occupier has no proper claim, when the horse is reclaimed by someone entitled to its possession.

4 If a horse has been detained for at least 14 days the occupier may sell it at market or by auction, unless any court proceedings are pending. The occupier is entitled to the proceeds of the sale, but any excess over his claim has to be paid to the person who but for the sale would have been entitled to possession.

5 Occupiers are liable for any damage caused to the horse that they have detained, by failing to treat it with care, or by failing to supply it with food and water.

6 Occupiers who sell the animal have no right to sue for damages. They must choose one remedy or the other.

Fencing

The Act does not define a 'duty to fence', so common-law rules have to be considered. These show that the law has never imposed any obligation on a person to fence his land, but obligations may have been created by formal agreements or may have become established through long usage or custom. For example, if a person has maintained a fence for, say, forty years and upwards this would indicate that he thought he had a duty to do so. He cannot then deny the obligation, and his actions have created a liability on him and his successors in title.

Where horses stray from common land where they had a 'right' to graze, the position might differ if it can be established that it was the responsibility of the land owner to fence against animals straying from it.

Any fences erected must be kept in a proper state of repair so as not to constitute a danger to adjoining occupiers or to people on a highway.

Dogs

This Act also gives protection against dogs either killing or causing injury to livestock. For proceedings and defences, reference should be made to Section 9 of the Act.

PROTECTION OF ANIMALS ACT 1911—ENGLAND AND WALES

PROTECTION OF ANIMALS ACT (SCOTLAND) 1912

These Acts are perhaps the most widely used of the many Welfare of Animals Acts currently on the Statute Book. Both Acts are similar in content. Any reference to magistrates in the *Protection of Animals Act 1911* should be read as a reference to the sheriff in the *Protection of Animals Act (Scotland) 1912*. As the powers of courts in regard to penalties are subject to changes under the Criminal Justices Act, these have not been included.

The Acts consolidate the laws relating to cruelty to domestic and captive animals. The expression 'domestic animal' covers any animal of whatever species which is tame, or which has been, or is being sufficiently tamed to serve some purpose for the use of man. 'Captive animal' means any animal which is in captivity or confinement.

The 1911 Act has 15 sections but this chapter only deals with those sections which are relevant to horses. The expression 'horse' includes mare, gelding, pony, foal, colt, filly, stallion, ass, mule or jennet.

The Offences of Cruelty in Section 1 Read that if any Person:

(a)   'Shall cruelly beat, kick, ill-treat, over-ride, over-drive, over-load, torture, infuriate, or terrify any animal, or shall cause or procure, or, being the owner, permit any animal to be so used, or shall, by wantonly or unreasonably doing or omitting to do any Act, or by causing or procuring the commission or omission of any Act, cause any unnecessary suffering, or being the owner permit any unnecessary suffering to be so caused to any animal, shall be guilty of an offence.'

This long paragraph indicates, among many other things,

that if a person carries out any of the above offences, or hires or permits any other person to commit them, it will be an offence. It also means that by unreasonably preventing any of the following from taking place, the owner and/or offender will be guilty within the meaning of the Act. Failure to feed, failure to provide veterinary treatment, or failure to slaughter an animal which is incurably diseased and thereby causing unnecessary suffering may well be offences of omission. The abandonment of animals was not an offence under this Act as it was originally passed, so the *Abandonment of Animals Act 1960* was passed to rectify the omission. It is now an offence to abandon an animal (whether permanently or not) in circumstances likely to cause that animal any unnecessary suffering. This would appear to include both liberating animals in order to get rid of them, and leaving animals shut up unattended.

The clause of causing or procuring the commission of certain acts by others, could be interpreted that the parents of children who, acting on the instructions of their parents, perform acts of cruelty, could be guilty. In such a case it would be irrelevant if the child is below the age of criminal responsibility. There is no reason why one person should not be charged with 'causing' or presumably 'permitting' and another with ill-treating in respect of the same act. This would be the case where a horse is ridden in an unfit state (e.g. lame), the stable manager (or the owner) knowing/causing (or permitting) it to be ridden, and the rider for knowingly riding it.

CRUELTY AND UNNECESSARY SUFFERING. The mere infliction of pain is not sufficient to constitute any offences given in the Act. It has been ruled that cruelty is defined as: 'causing unnecessary suffering', and courts have to examine whether the defendant is doing something which it was not reasonably necessary to do. If the reason was of sufficient importance to justify the act done then no offence has been committed. Suffering must not only be dispropor-tionate to the alleged reason for which it was inflicted, but must also be substantial. What is clear from judgments given is that 'cruelly' and 'so as to cause unnecessary suffering' mean exactly the same thing. The inflicting of mental suffering by the defendant's own positive act could be within the category of cruelly infuriating and terrifying animals; it may also be included in the category of torturing animals.

(b)   'Shall convey or carry or cause or procure, or, being the owner permit to be conveyed or carried, any animal in such a manner or position as to cause that animal any unnecessary suffering'.

This sub-paragraph refers to all kinds of animals carried, and in regard to horses it is generally assumed to be by mechanical methods. Other legislation exists where offences of carrying can occur both by air or sea, but perhaps the most relevant legislation is the *Transit of Animals (Road and Rail) Order* of 1975 and amendments 1979.

(d)   'Shall wilfully, without any reasonable cause or excuse, administer, or cause or procure or being the owner permit such administration of, any poisonous or injurious drug or substances to any animal, or shall wilfully without any reason cause or excuse, cause such substances to be taken by any animals'.

This means that the administration of any injurious drug which causes unnecessary suffering, or the owner attempting or permitting an attempt by another person, to poison an animal is regarded as an offence.

(e)   'Shall subject or cause or procure, or being the owner permit, to be subjected, any animal to any operation which is performed without due care and humanity'.

Treatments and operations to agricultural animals (as defined in the *Agriculture Act 1947*) which are permitted to be carried out by unqualified persons, are specified in Schedule III of the *Veterinary Surgeons Act 1966*.

The Powers of a Court to Order Destruction of Animals in Section 2

Where the owner of an animal is convicted of an offence of cruelty within the meaning of this Act, the court has powers, providing they are satisfied that it would be cruel to keep the animal alive, to direct that: (a) the animal should be destroyed, and (b) the animal should be assigned to any suitable person for destruction, and the person to whom it is assigned shall as soon as it is possible destroy that animal or arrange for its destruction in his/her presence without unnecessary suffering. The court may order that any reasonable expenses incurred in the destruction shall be paid by the owner of the animal or be recoverable by a summons as a civil debt. It should be noted that unless the owner agrees to the destruction, a veterinary surgeon must give evidence to the court to the effect that it would be cruel to keep the animal alive. If no such consent, or evidence, is given, courts cannot make an order for destruction.

There can be no appeal against an order for the destruction of an animal. However, the magistrates may themselves suspend or rescind an order on complaint.

The Powers of Confiscation in Section 3

If the owner of any animal shall be guilty of cruelty within the meaning of this Act, the courts, upon conviction of the owner, may if they think fit, in addition to any other punishment, deprive such a person of the ownership of the animal, and they may also make an order as to the disposal of the animal as they think fit under the circumstances. They are not permitted to make such an order, unless it is shown by evidence of a previous conviction, or as to the character of the owner, or otherwise, that if the animal is left with the owner, it is likely to be exposed to further cruelty.

The Regulations for Knackers in Section 5

Any police constable has a right of entry to a knacker's yard during hours when business is usually being carried out, for the purpose of examining if there has been any contravention or non-compliance with the provisions of this Act. For the purposes of Section 1 of this Act relating to offences of cruelty the knacker shall be regarded as the owner of any animal delivered to him. The expression 'knacker' means a person whose trade or business is to kill any cattle including horses for the purpose of the flesh not being used as butchers' meat.

Animal Pounds in Section 7

> 'If someone personally, or causes, any animal to be impounded or confined in any pound, they shall while the animal is so impounded or confined, be responsible for supplying a sufficient quantity of wholesome and suitable food and water to that animal, and if they fail to do so will be liable upon conviction to a fine.'

The person taking the animal to the pound is therefore responsible for it under this Section, not the pound keeper. The pound keeper may be guilty of an offence under sections of this Act if it can be shown that he was legally responsible for performing that act.

If any animal impounded or confined is without sufficient suitable food or water for six successive hours, or longer, any person may enter the pound for the purpose of supplying the animal. The reasonable cost of the food and water supply shall be recoverable from the owner of the animal as a civil debt.

Poisoned Animal Food in Section 8

> 'It is an offence to offer for sale, to give away or to cause or procure any other person to do so any grain or seed which has been rendered poisonous, except for bona-fide use in agriculture. It is also an offence to place, cause some other party to place, or knowingly be a party to the placing in or upon any land or building any poison or fluid or edible matter (which is not sown seed or grain) that has been rendered poisonous.'

The defence can be used that the poison was placed by the accused for the purpose of destroying insects or small ground-vermin where such is necessary in the interest of public health, agriculture or the preservation of other animals, or for the purpose of manuring the land, providing that all reasonable precautions to prevent injury to all domestic animals and wild birds were taken.

Diseased and Injured Animals in Section 11

(i) If a Police Constable finds any animal so distressed or so severely injured, or in such a physical condition that in his opinion, there is no possibility of removing it without cruelty, he shall in the absence of the owner, or if the owner refuses to consent to the animal being destroyed, summon a Veterinary Surgeon if any reside within a reasonable distance and providing it appears by the certificate of that Veterinary Surgeon that the animal is mortally injured, or so severely injured, or so diseased, or in such a physical condition that it is cruel to keep it alive, it is lawful for the Police Constable, without the owner's consent to slaughter or arrange for the slaughter, with such instruments or appliances, and with such precautions, and in such a manner as to inflict as little suffering as is possible. The Constable also has powers to arrange for the removal of the carcasses if the slaughter of the animal takes place on a Public Highway.

(ii) If a Veterinary Surgeon certifies that the animal can be removed without cruelty, it is the responsibility of the person in charge to arrange for the removal with as little suffering as possible, and if the person in charge fails to do this, the Police Constable may without the consent of the person in charge cause the animal to be removed.

(iii) Any reasonable expenses incurred including those of the Veterinary Surgeon and regardless of whether the animal is slaughtered or not may be recovered from the owner as a civil debt.

(iv) 'Animal' in this section applies only to horses (as defined in the preamble), and to bull, sheep, goat or pig.

The Summons in Section 13

Frequently, the most valuable piece of evidence is the animal itself. If the owner refuses to produce the animal, or to have it examined by a veterinary surgeon, a summons may be applied for ordering the production of the animal in the same way as a summons to an accused person.

Rights of Appeal in Section 14

(i)　　　　Any appeal from conviction or order (other than appeal against destruction in Section 2) made by a summary court, shall be made to a Crown Court.

(ii)　　　Where there is an appeal against a summary court conviction or order, the court may order the owner not to sell or part with the animal until the appeal is determined or abandoned, the court may order that the animal be produced on the hearing of the appeal providing such production is possible without cruelty being incurred.

PROTECTION OF ANIMALS (AMENDMENT) ACT 1954

This Act, which applies to Scotland but not to Northern Ireland, extends the powers of the court to disqualify from having custody of animals, a person previously convicted of an offence of cruelty to an animal.

Powers of Disqualification in Section 1

(i)　　　　Where a person has been convicted under the Protection of Animals Act 1911 or the Protection of Animals (Scotland) (1912), and is subsequently convicted under either of these Acts of an offence, the court by which the subsequent conviction is made, may if it thinks fit either, in addition to, or in substitution for, any other punishment, order the person to be disqualified, for a such period as it thinks fit, from having custody of any animals or any animal of a kind specified.

(ii)   A court which has made such a disqualification order may if it thinks fit suspend the order:
(a) For a period of time as the court thinks necessary to enable arrangements to be made for the custody of any animals to which the order relates; or
(b) Pending an appeal.

(iii)   Any disqualified person may after twelve months from the date of the order and subsequently from time to time, apply to the court by which the order was made for removal of the disqualification, and the court will consider the character of the applicant, the conduct following the order, the nature of the offence to which he was convicted, and any other circumstances of the case; and either:
(a) Direct the removal of the disqualification from such a date as they specify, or shall vary the order to apply only to animals of a kind specified; or
(b) Refuse the application.

Further applications for variations cannot be made within twelve months after the date of the direction or refusal.

ADDITIONAL EFFECTS OF DISQUALIFICATION. Any person disqualified under this Act from having custody of an animal may not be granted a licence to keep either an animal-boarding establishment or a riding establishment, a pet shop, or dog-breeding kennels.

Further penalties can be imposed for breaches of any Disqualification Order.

PROTECTION OF ANIMALS (AMENDMENT) ACT 1988
Gives power to the court to disqualify a person convicted of cruelty to animals from having custody of animals.

PROTECTION AGAINST CRUEL TETHERING ACT 1988
It shall be an offence to tether any horses, asses or mules under such conditions or in such manner as to cause that animal unnecessary suffering.

121

Bibliography

ABRS, *The Law on the Buying and Selling of Ponies.*

ABRS, *Negligence in Riding Schools.*

BRITISH HORSE SOCIETY, Pamphlets on *Building an Outdoor Arena; Buying Ponies; Horse Welfare.*

BRITISH HORSE SOCIETY, *The Manual of Horsemanship* (Threshold Books).

HEALTH AND SAFETY EXECUTIVE, *Safety* pamphlet.

HICKMAN, PROFESSOR, FRCVS, *Stable Management* (Academic Press).

HMSO, *Animal Management.*

HORSE AND HOUND Survey (IPC Magazines).

HOUGHTON-BROWN, J., and POWELL-SMITH, DR V., *Horse and Stable Management* (Granada Publishers Ltd).

MACDONALD, JANET, *Running Stables as a Business* (J.A. Allen).

ROSE, MARY, FBHS, *The Horsemaster's Notebook* (Threshold Books).

SMITH, PETER, ARIBA, *Design and Construction of Stables* (J.A. Allen).

SOPHIAN, *Horses and the Law* (J.A. Allen).

THOMSON, BILL, *Constructing Cross-Country Obstacles* (J.A. Allen).

Index